INTELLIGENT ROBOTIC SYSTEMS:

THEORY, DESIGN AND APPLICATIONS

THE KLUWER INTERNATIONAL SERIES IN ENGINEERING AND COMPUTER SCIENCE

ROBOTICS: VISION, MANIPULATION AND SENSORS

Consulting Editor: **Takeo Kanade**

ROBOT DYNAMICS ALGORITHMS, F. Featherstone
ISBN: 0-89838-230-0
KINEMATIC MODELING, IDENTIFICATION AND CONTROL OF ROBOT MANIPULATORS, H.W. Stone
ISBN: 0-89838-237-8
OBJECT RECOGNITION USING VISION AND TOUCH, P. Allen
ISBN: 0-89838-245-9
INTEGRATION, COORDINATION AND CONTROL OF MULTI-SENSOR ROBOT SYSTEMS, H.F. Durrant-Whyte
ISBN: 0-89838-247-5
MOTION UNDERSTANDING: Robot and Human Vision, W.N. Martin and J. K. Aggrawal (editors)
ISBN: 0-89838-258-0
BAYESIAN MODELING OF UNCERTAINTY IN LOW-LEVEL VISION, R. Szeliski
ISBN 0-7923-9039-3
VISION AND NAVIGATION: THE CMU NAVLAB, C. Thorpe (editor)
ISBN 0-7923-9068-7
TASK-DIRECTED SENSOR FUSION AND PLANNING: A Computational Approach, G. D. Hager
ISBN: 0-7923-9108-X
COMPUTER ANALYSIS OF VISUAL TEXTURES, F. Tomita and S. Tsuji
ISBN: 0-7923-9114-4
DATA FUSION FOR SENSORY INFORMATION PROCESSING SYSTEMS, J. Clark and A. Yuille
ISBN: 0-7923-9120-9
PARALLEL ARCHITECTURES AND PARALLEL ALGORITHMS FOR INTEGRATED VISION SYSTEMS, A.N. Choudhary, J. H. Patel
ISBN: 0-7923-9078-4
ROBOT MOTION PLANNING, J. Latombe
ISBN: 0-7923-9129-2
DYNAMIC ANALYSIS OF ROBOT MANIPULATORS: A Cartesian Tensor Approach, C.A Balafoutis, R.V. Patel
ISBN: 07923-9145-4
PERTURBATION TECHNIQUES FOR FLEXIBLE MANIPULATORS: A. Fraser and R. W. Daniel
ISBN: 0-7923-9162-4
COMPUTER AIDED MECHANICAL ASSEMBLY PLANNING: L. Homen de Mello and S. Lee ISBN: 0-7923-9205-1
INTELLIGENT ROBOTIC SYSTEMS FOR SPACE EXPLORATION: Alan A. Desrochers
ISBN: 0-7923-9197-7
MEASUREMENT OF IMAGE VELOCITY: David J. Fleet
ISBN: 0-7923-9198-5
DIRECTED SONAR SENSING FOR MOBILE ROBOT NAVIGATION: John J. Leonard, Hugh F. Durrant-Whyte ISBN: 0-7923-9242-6
A GENERAL MODEL OF LEGGED LOCOMOTION ON NATURAL TERRAIN: David J. Manko
ISBN: 0-7923-9247-7

INTELLIGENT ROBOTIC SYSTEMS:

THEORY, DESIGN AND APPLICATIONS

By:

Kimon P. Valavanis
Associate Professor of Computer Engineering
Intelligent Robotic Systems Laboratory,
The Center for Advanced Computer Studies,
The University of Southwestern Louisiana,
Lafayette, LA 70504-4330, USA

George N. Saridis
Professor of Electrical, Computer and Systems Engineering
Director, NASA Center for Intelligent Robotic Systems
for Space Exploration,
Rensselaer Polytechnic Institute,
Troy, NY 12180-3590, USA

KLUWER ACADEMIC PUBLISHERS
Boston/ London/Dordrecht

Distributors for North America:
Kluwer Academic Publishers
101 Philip Drive
Assinippi Park
Norwell, Massachusetts 02061 USA

Distributors for all other countries:
Kluwer Academic Publishers Group
Distribution Centre
Post Office Box 322
3300 AH Dordrecht, THE NETHERLANDS

Library of Congress Cataloging-in-Publication Data

Valavanis, K. (Kimon)
Intelligent robotic systems : theory, design, and applications / by Kimon P. Valavanis, George N. Saridis.
p. cm. — (The Kluwer international series in engineering and computer science ; SECS 182)
Includes bibliographical references and index.
ISBN 0-7923-9250-7
1. Robotics. I. Saridis, George N., 1931- . II. Title. III. Series.
TJ211.V35 1992
629.8'92—dc20 92-10876
CIP

Printed on acid-free paper.

Printed in the United States of America

for

Areti, little Stella (KPV)

and

Youla (GNS)

Contents

LIST OF FIGURES

LIST OF TABLES

Preface

The word robot was derived from the Czech word *robota* (work) and was first used by Karel Capek, a Czech writer, in his play Rossum's Universal Robots (R.U.R.) in 1921, to describe machines created to replace humans. The official definition for a robot given by R.I.A. (Robot Institute of America) considers a robot to be *"a programmable, multifunction manipulator designed to move material, parts, tools, or specific devices through variable programmed motions for the performance of a variety of tasks."* In that respect, every automated industrial machine may be considered as a primitive robot.

Since the late 1960s, there has been a revolution in robots and industrial automation, from the design of robots with no computing or sensory capabilities (first-generation), to the design of robots with limited computational power and feedback capabilities (second-generation), and the design of *intelligent* robots (third-generation), which possess diverse sensing and decision making capabilities.

The development of the Theory of Intelligent Machines has been developed in parallel to the advances in robot design. This Theory is the natural outcome of research and development in Classical Control (1950s), Adaptive and Learning Control (1960s), Self-Organizing Control (1970s) and Intelligent Control Systems (1980s).

The Theory of Intelligent Machines involves utilization and integration of concepts and ideas from the diverse disciplines of Science, Engineering and Mathematics, and fields like Artificial Intelligence, System Theory and Operations Research. The main focus and motivation is to bridge the gap between the diverse disciplines involved and bring under a common cover several generic methodologies pertaining to what has been defined as *machine intelligence.*

Intelligent Robotic Systems are a specific application of Intelligent Machines. They are complex computer controlled robotic systems equipped with a diverse set of visual and non visual sensors and possess decision making and problem solving capabilities within their domain of operation. Their modeling and control is accomplished via analytical and heuristic methodologies and techniques pertaining to Generalized System Theory and Artificial Intelligence.

The work reported in this monograph was originated by the authors in 1982 at the Robotics and Automation Laboratory of Rensselaer Polytechnic Institute. The authors' long term cooperation has produced this monograph which represents a considerable amount of work during the last seven years. It presents and justifies the fundamental concepts and ideas associated with the modeling and analysis of Intelligent Robotic Systems. This research has been motivated by: i) Ilya Prigogine's philosophical and mathematical approach related to the *physics of being*, the *physics of becoming* and *the bridge from being to becoming* in dynamical systems, the "state" of which is measured with entropy functions, ii) the Generalized System Theory of Conant who has provided laws of information which govern the performance of dynamical systems based on entropy and transmission rates, and, iii) Boltzmann's Theory of Statistical Thermodynamics where entropy is used as an energy function.

The proposed analytical design for Intelligent Robotic Systems is based on the duality of the entropy function as a measure of uncertainty (Shannon's definition) and as an energy function (Statistical Thermodynamics), utilized to unify the performance of such systems. The system model follows the principle of Increasing Precision with Decreasing Intelligence, and is composed of three hierarchical interactive levels of organization, coordination and execution of tasks. A different type of intelligence, machine intelligence, is introduced. A bottom-up synthesis approach, similar to a linguistics paradigm where a language is generated from 26 letters, is followed for plans formulation and organization. Plans are formulated from a set of primitive events/actions and checked for compatibility and completeness. A unique architecture is designed to accommodate fast and reliable operation of the Intelligent Robotic System.

This monograph is appropriate for researchers and engineers in the area of robotics and automation. It may be used as a text in a graduate level course in Intelligent Robotics/Machines, as well as a reference book.

The authors wish to acknowledge their colleagues for their support and encouragement during the period this research was conducted and the monograph was written. Dr. A. Meystel was the first who encouraged us to write this monograph. Special thanks go to Drs. M. Eleftheriou, K. Kyriakopoulos, M. Leahy, S. Murphy, C. Seaman, F. Wang, all members of the RPI research "gang" of the early and middle 1980s. Our research cooperation and social activities have been memorable. Special thanks also go to Dr. Alex Levis who introduced us to his theory of organizations. The first author would also like to acknowledge Drs. C. Jacobson and P. Serafim, former colleagues from Northeastern University, and his current colleagues, faculty and staff, at the Center for Advanced Computer Studies (especially Drs. M. A. Bayoumi and M. C. Mulder) for making this first year at The Center very enjoyable. The second author would like to acknowledge the late Prof. K. S. Fu whose inspiration and friendship produced the seeds of the Theory of Intelligent Controls. The authors are also grateful to Mrs. Joan Heine for the excellent and professional preparation of the monograph, P. Landry and M. Radle (our computer experts) and Mr. S. Malladi for the preparation of the figures. The editor, Mr. Thomas Stone from Kluwer has been very cooperative, supportive, patient and enthusiastic during the preparation of this document.

This research work has been partially supported by the RPI NASA/CIRSSE USERC Program NAGW 1333 Research Grant and the Louisiana Educational Quality Support Fund Research Grant LEQSF (1991–94) RD-A-43.

Kimon P. Valavanis and George N. Saridis

Chapter 1
INTRODUCTION

1.1 Introduction

Systems operating autonomously or semi autonomously in structured or unstructured environments may be driven by controls with special characteristics described by the methods of Hierarchically Intelligent Control Systems [185].

Intelligent control is the discipline that implements Intelligent Machines (IMs) to perform anthropomorphic tasks with minimum supervision and interaction with a human operator. It may involve elements of three different disciplines: Artificial Intelligence, Operations Research and System Theory, as shown in Figure 1.1.

Intelligent Machines are defined to be hierarchical structures in the order of increasing intelligence and decreasing precision. This intelligence, called machine intelligence (as it will be justified in the sequel), is embedded in the hierarchical structure of the system.

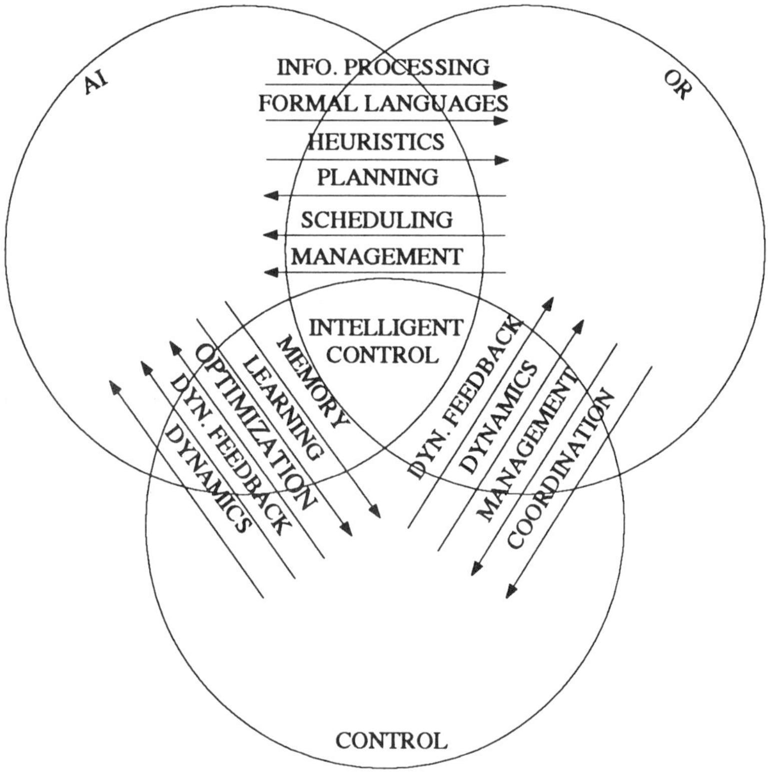

Figure 1.1 Definition of the Intelligent Control Discipline

Intelligent Machines differ from other hierarchical structures described in Findeisen, Baily, et al [59], in this particular type of ordering, putting emphasis on the intelligence of the machine.

Artificial Intelligence (AI) has been used in knowledge based systems as an interactive means of gathering and discriminating information with human operators. It originated with the idea to simulate (somehow) most of the functions of the human brain on a computer, such that scientists may understand the functions of the human brain. Expert systems were the natural applications of such an activity. In order to design Intelligent Machines to perform anthropomorphic tasks in familiar or unfamiliar but uncertain environments, one should expect to utilize the abilities of modern computers and robot systems to their maximum. This requires a different type of intelligence, the intelligence built in the machine called Machine Intelligence (MI). Machine Intelligence maps the abilities of a machine into the space of anthropomorphic tasks it is asked to perform with minimum interaction with a human operator. Machine Intelligence is the internal function and mechanism of the system which produces enhanced performance, generates and chooses from a set of alternative actions based on accumulated knowledge (information) from a diverse set of sensors interacting with the environment. If Artificial Intelligence represents the mapping of anthropomorphic functions on the computer, Machine Intelligence represents the inverse mapping which may not be one to one.

Most attempts made to create Intelligent Machines are based on digital computers which utilize human logic and heuristics in an effort to imitate the function and the activities of the human brain [40, 81, 82, 105]. In spite of the tremendous capacity of the modern computers to heuristically organize and execute complex tasks, it is highly desirable to generate an analytic approach to provide a means to optimize the operation and thus produce much more efficient machines.

In an attempt to formulate an analytic approach to design Intelligent Machines one may first consider the diversity of functions involved in such an operation, from the high level computer organization and coordination of tasks to their hardware execution, along with their cross disciplinary interaction [184, 186, 190].

A major problem associated with such designs is the uncertainties in the environment of operation of Intelligent Machines and the internal decision making procedures within the machines. These uncertainties are mostly caused by poor or inefficient modeling of the environment of operation. They may be treated and overcome using either heuristic or analytical methodologies, or a combination of both. Aprobabilistic approach (two value logic) assigns a probability density function (pdf) to describe how uncertain is the outcome of a particular event selected from a class of many possible events [27, 36, 122, 231, 237]. The appropriate entropy function may serve as a measure of this uncertainty. A Fuzzy Set Theory based approach (multi value logic) assigns a membership function instead of a probability density function to achieve similar goals [223]. Analytic approaches based on generalized N-dimensional Information Theory utilize entropy functions and their rates to define activity rates within the system and minimize all related uncertainties by minimizing entropy based performance indices [12, 32, 34, 43–46, 79, 232, 242]. Other modeling approaches include knowledge based expert systems, formal languages, linguistic and connectionist models and different types and classes of Petri net (PN) models.

This monograph deals with a special application of Intelligent Machines: the Intelligent Robotic Systems (IRSs). All models and algorithms derived in the monograph are based on the Hierarchically Intelligent Control Systems Theory designed to implement Intelligent Machines, originated by Saridis and Stephanou [187] and established by Saridis [184–202], and the Generalized System Theory of Conant [43–46]. Furthermore, the duality of the concept of entropy as defined in Information Theory [71] and Theoretical Thermodynamics [122] may be utilized to create measures for the analytical design of Intelligent Machines operating under the constraints of Hierarchically Intelligent Control Systems Theory. Hierarchically intelligent control applies a mathematical approach for problem solving and job execution. A three interactive level structure is proposed herein, based on the principle of Increasing Precision with Decreasing Intelligence (IPDI) which will be explained in chapter 3, and used to accommodate fast and reliable operation of the functions of the Intelligent Machine. This is useful because it integrates the functions of the computer logic, the sensory devices and the hardware control into one system. If all the levels of the hierarchically intelligent system are described by mathematical models, then the optimal operation of the system may be obtained as the solution of the following mathematical programming problem: *Intelligent control is postulated as the mathematical problem of finding the right sequence of internal decisions and controls for a system structured according to the principle of Increasing Precision with Decreasing Intelligence such that it minimizes its total entropy* [190].

Since Intelligent Robotic Systems are a special application of Intelligent Machines, from now on both terms will be used interchangeably. Although this is the case, chapter 3 justifies how the general theory of Intelligent Machines is applied to Intelligent Robotic Systems. Furthermore, all case studies in chapter 6 are related to Intelligent Robotic Systems.

A general purpose Intelligent Robotic System considered in this monograph is composed of:

1. A high level digital computer capable of responding to fuzzy or qualitative commands, required for the intelligent operations of the system.
2. General purpose manipulators (mobile or not, redundant or not) with a gripper (or a set of grippers) which opens and closes.
3. Manipulator controllers (one for each manipulator) with dedicated microcomputers to control directly the associated robot manipulator.
4. A vision system with camera(s) and laser range finders that recognizes various objects and provides visual information about their position and orientation, and,

5. A sensory system with touch, force/torque, proximity and tactile sensors that provides all kinds of sensory (non visual) information.

Coordination of the different subsystems provides the means to integrate all individual functions performed by each subsystem into one complex system.

Regardless of the point of view and specific approach used, desirable features of any related modeling methodology associated with such complex systems should consider:

1. System reliability, stability and controllability.
2. System on line and off line learning capabilities.
3. System adaptability to unforeseen events, and,
4. System ability to recover from local failures.

The general structure of a hierarchically intelligent system is shown in Figure 1.2. A general block diagram of an Intelligent Machine is shown in Figure 1.3. The configuration of an Intelligent Robotic System with multi sensory capabilities is shown in Figure 1.4. Explanations and justifications for the functions within each block are given in subsequent chapters.

The proposed hierarchy encaptures both *i) fixed structure multi level control, ii) multi layer control* nested in the first one [119]. The overall system is physically decomposed into subsystems controlled by independent controllers, with the coordination level ensuring that all interactions are taken into account. Aggregated information is used to define a decision making strategy based on the principle of Increasing Precision with Decreasing Intelligence (multi layer control) [196, 232, 237].

It is very important to clarify that the *number of levels is crucial since each level functions with clearly defined objectives.* Each level may be decomposed into several layers depending on the complexity of the task to be executed. The proposed three level hierarchy has been followed and used in a diverse class of problems.

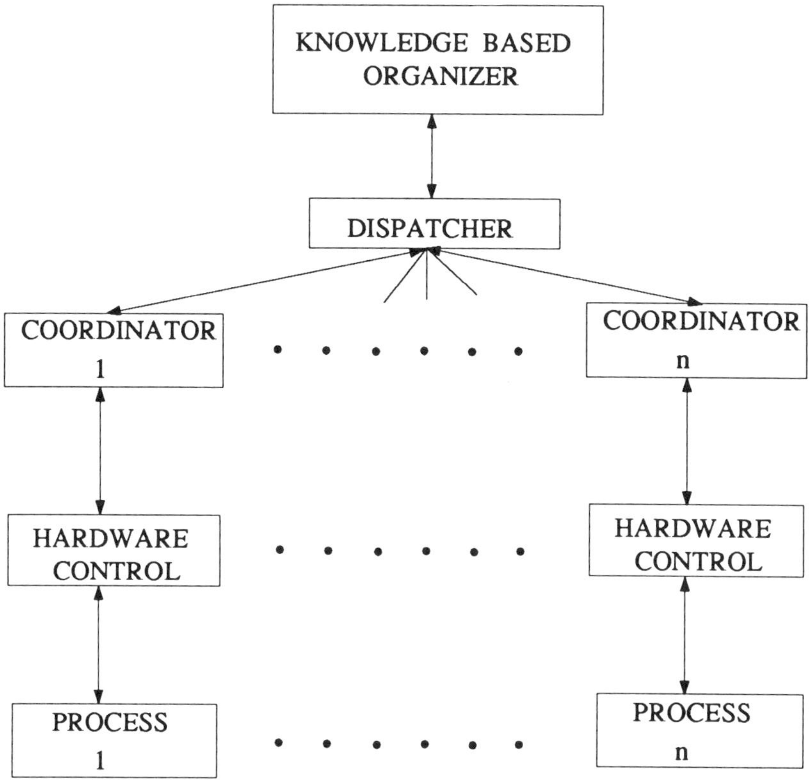

Figure 1.2 General Structure of a Hierarchically Intelligent Control System

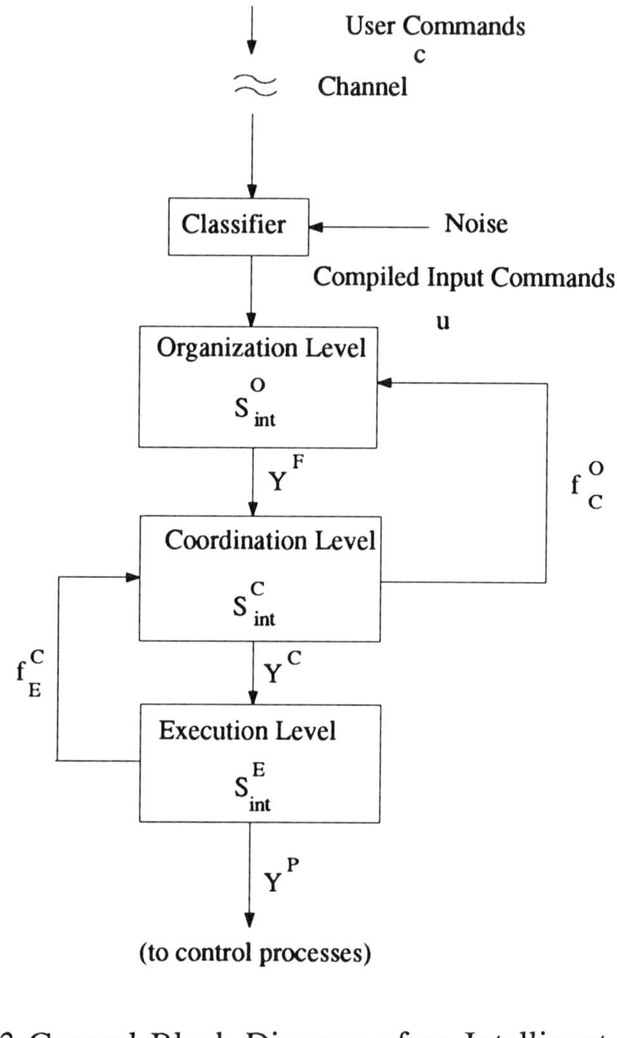

Figure 1.3 General Block Diagram of an Intelligent Machine

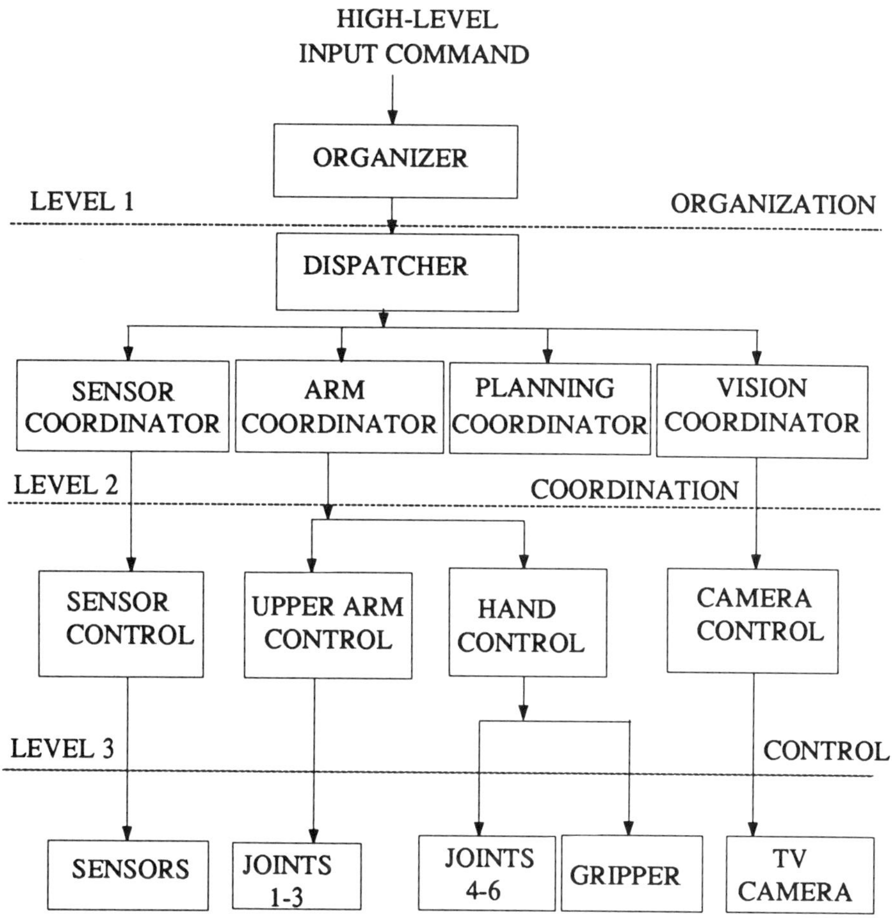

Figure 1.4 Intelligent Robotic System with Multi-Sensory Capabilities

The three levels are: i) the organization level, ii) the coordination level, and iii) the execution level, as shown in Figures 1.3 and 1.4. Their functions as originally developed, and modified later on, are explained in detail in chapter 3, while the corresponding models are presented in chapter 4. A brief summary of the objective of each level follows.

The organization level is basically structured after a knowledge based system [40, 81, 82]. Two different models are discussed: a probabilistic model (binary logic) and an expert system model. Concepts and ideas from Artificial Intelligence are mathematically expressed to represent the organizer functions. Forward and backward chaining is utilized to generate plan scenarios.

The coordination level is composed of a number of coordinators supervised by a dispatcher. The structure of each coordinator is fixed. Each coordinator is designed to perform its own individual functions. Communication between the different coordinators is performed through a dispatcher whose variable configuration is dictated by the organizer. Feedback information selectively received from the execution level allows the coordination level to modify on line execution scenarios. Selective feedback from this level is also communicated to the organizer after the execution of the requested job (off line).

The execution level is composed of hardware controllers that drive the mechanical areas of the system, the image and sensory information processing systems. Specific execution devices are associated with each coordinator. They execute specific functions when the corresponding coordinator is accessed.

Each level is modeled in such a way that it evaluates the performance of the immediately lower one through learning algorithms. The overall system structure resembles that of a loosely coupled parallel processing system.

1.2 Outline of the Monograph

This monograph consists of seven chapters. Chapter 2 presents a well balanced and thorough review of models and methods used to model hierarchical and knowledge based (expert) systems in general. Since most of the initial work in this area has been based on Information Theory, the review includes a summary of such related approaches. More recent methodologies on intelligent controls and different schools of thought are also included. A summary of knowledge based systems along with some comparison between them and Intelligent Machines concludes the chapter. Chapter 3 presents the general theory of Intelligent Machines. The main objective of this chapter is to set the theoretical foundations upon which the actual modeling and design of the machine is based. Several justifications are provided to clarify new concepts like the principle of Increasing Precision with Decreasing Intelligence and the flow of knowledge within the Intelligent Machine. Chapter 4 discusses the models for the three levels, while chapter 5 deals with the proposed hardware architectures. Specific hardware mechanisms related to the system individual functions are presented. Three case studies are discussed in chapter 6. The last chapter summarizes the monograph, discusses very recent developments in this area and suggests potential future research directions. An extended list of references concludes this monograph. The reference list is by no means complete nor it favors certain approaches over others. Although we are aware of the voluminous amount of work in this area, the reference list reflects the research work pertaining to the preparation of this monograph.

Chapter 2
A REVIEW ON HIERARCHICAL AND EXPERT SYSTEMS

2.1 Introduction

The problem of designing Intelligent Machines belongs to more than one conventional research area. In order to provide a solid framework upon which the analytical design will be based, ideas, concepts and measures from several branches of engineering, science and mathematics have been utilized and put together. This chapter provides a well balanced review of the evolution of pertinent methodologies and approaches for the design of Intelligent Machines as originally conceived. Applications of Information Theory to modeling of decision makers, general control and communication systems are reviewed first, followed by a summary of hierarchically intelligent system methodologies and a short review of knowledge based expert systems.

2.2 Information Theory as Applied to General Systems Theory and Modeling of Decision Makers

Information Theory first introduced by Shannon [207, 208], studies the communication of messages from one point to another, and its focus has remained on the question of how the constraint between two variables X and Y that respectively represent a message sent and a message received can be measured and maximized, in a channel with limited capacity.

The basic measure of Information Theory, associated with a discrete random variable X (which may also be the discretized version of a continuous variable) is the entropy $H(X)$, a function of the probability distribution of X, defined as follows:

$$H(X) = -\sum_{x} p(x) \log p(x) \tag{2.2.1}$$

The summation is over all values x taken by X, $p(x)$ is the probability density function of X while the base of the logarithm is specified accordingly.

The entropy $H(X)$ is the average uncertainty of particular values x taken by X, where uncertainty is measured by $-\log p(x)$. It is the average amount of information obtained, given the value x taken by X on an arbitrary occasion if the distribution *p(x)* is known a priori. Conditional entropies are defined through conditional or joint distributions as:

$$H_Y(X) = -\sum_{x,y} p(x,y) \log p(x/y) = -\sum_{y} p(y) \sum_{x} p(x/y) \log p(x/y) \tag{2.2.2}$$

where *p(x,y)* is the joint probability density function and *p(x/y)* the conditional one. Conditional entropies denote the uncertainty remaining in one variable after knowledge of the other.

In a similar way, the differential entropy $H(X)$ of a continuous random variable X with density function $f(x)$ (with respect to

Lebesque measure), is defined by Van Campenhout and Cover [244] as:

$$H(X) = -\int_{-\infty}^{+\infty} f(x)\, ln f(x) dx \tag{2.2.3}$$

with the corresponding terms defined accordingly. Conditional and joint differential entropies are defined similarly.

The amount by which two random variables are related, is measured by the transmission of information between them, denoted $T(X : Y)$. Transmission of information is defined either through individual and joint entropies, or, through conditional entropies as shown below:

$$T(X : Y) = H(X) + H(Y) - H(X, Y) \tag{2.2.4a}$$

$$T(X : Y) = H(X) - H_Y(X) \tag{2.2.4b}$$

$$T(X : Y) = H(Y) - H_X(Y) \tag{2.2.4c}$$

The transmission of information measures the relation between variables, or, the amount by which knowledge of one variable reduces the uncertainty about the other. Conditional transmission is defined similarly.

The entropy function possesses *additivity properties* which allow the partitioning of the (overall) uncertainty of several variables, into the uncertainty of the first plus the uncertainty of the second remaining after knowledge of the first, etc. Transmissions and conditional transmissions are non negative, like entropies and conditional entropies, which, with their additivity properties allow the total relatedness or constraint between groups of variables to be partitioned in a manner similar to the partition of entropy.

Entropy and transmission rates are defined to understand the overall behavior of systems whose past history influences their present value to a large extent. The entropy rate is defined by Conant as the information carried per observation in a long sequence $< X(t), X(t+1), ..., X(t+m-1) >$ with units in bits per step [44]:

$$\bar{H}(x) = \lim_{m \to \infty} (1/m) H(X(t), X(t+1), ..., X(t+m-1)) \quad (2.2.5)$$

Hall and Levis [79], as well as Papoulis [165], define the entropy rate for systems with statistically independent inputs generated every τ seconds as:

$$\bar{H}(x) = (1/\tau) H(x) \quad (2.2.6)$$

Transmission rates are defined similarly. It will be assumed in this monograph that the entropy rate is given by (2.2.6). Transmission rates are then easily defined using (2.2.4) and (2.2.6).

There are many useful identities involving entropies and transmissions. Ashby has defined most of them [17]. Two of the most important identities are: the *rule of uniform subscripting* which states that if the same subscript is added to every quantity in an identity, the result is an identity, and the *rate rule* stating that if in an Information Theory identity every entropy and transmission is replaced by the corresponding rate, the result is another identity.

N-dimensional Information Theory has been used to design hierarchical systems because of three major advantages over other techniques of statistical analysis [43–46]:

1. Its ability to allow quantifications of the degree to which different parts of a system are in communication with each other.
2. Its ability to measure rates of constraint between dynamic variables with past history taken into account, and,
3. The additivity properties of the measures that allow decomposition of the constraints, so that the total constraint holding over all N variables of a system equals the constraint between the subsystems plus the sum of the constraints holding within each subsystem, the latter components being also decomposable.

Based on these advantages, researchers have attempted the derivation and explanation of the information laws which govern general dynamic systems, in addition to the evaluation of their performance in an information theoretic framework.

Conant [44], proposed measurement rates to study the behavior of dynamical systems. According to his approach a dynamical system is thought of as an ordered set of variables like:

$$G = < G_1, G_2, ..., G_n > = < G_j,\ j = 1, 2, ..., n > \quad (2.2.7)$$

The system receives a vector input E from its environment considered to be all relevant variables not in G. Those variables in G directly observable from its environment, constitute the output variables. The unobservable variables are internal variables. The above relation may be rewritten as:

$$G = < G_{int},\ G_{out} > \quad (2.2.8)$$

G_{out} is the subset of output (observable) variables and G_{int} is the subset of internal variables. Such a system is partitioned into N disjoint subsystems G_i of variables G_{ij}. The partition law of information rates (PLIR), derived by Conant, for the total (non deterministic) activity of such a system, F, if intervariable relationships are ignored is as follows:

$$F = F_t + F_b + F_c + F_n \quad (2.2.9)$$

$$F = \sum_j \bar{H}(G_j); \text{ the total rate of information flow} \quad (2.2.10)$$

$$F_t = \bar{T}(E : G_{out}); \text{ the throughput rate} \quad (2.2.11)$$

$$F_b = \bar{T}_{G_{out}}(E : G_{int}); \text{ the blockage rate} \quad (2.2.12)$$

$$F_c = \bar{T}(G_1 : G_2 : ... : G_n); \text{ the coordination rate} \quad (2.2.13)$$

$$F_n = \bar{H}_E(G); \text{ the noise rate} \tag{2.2.14}$$

The law simply states that the requirements on a system for selection of appropriate information and therefore blockage of irrelevant information, internal coordination of parts and throughput, are essentially additive and compete for the computational resources of the system. The simple additivity expressed in the law suggests the tradeoffs possible between them in a system of fixed total capacity. If estimates for blockage, coordination and output for a task are available and if the capacities of the system components are known, the law provides a lower bound on the number of components needed to accomplish the task. It is also shown that under certain assumptions systems are constraint losing as well as information losing.

Valavanis and Saridis [232] have extended the partition law of information rates and have derived a generalized partition law of information rates (GPLIR) for Intelligent Machines to accommodate all feedback information from the lower to the higher levels, internal control strategies and memory exchange procedures. The stochastic nature of the system was taken into account. Explicit expressions have been derived for the whole system, the organization and coordination levels.

In previous considerations of team theoretic decision problems [86–88], it has been assumed that the decision maker (DM) is perfectly rational, that is, each decision maker possesses a given set of alternatives, has knowledge in a probabilistic sense of the consequence of choosing a particular alternative, and has a cost ordering on the consequences [213]. As a result, optimal decision strategies were obtained. An alternative hypothesis however, was that due to limitations in information processing ability or problem solving ability, the decision maker is unable to construct and consider all alternatives in a given decision situation, nor can it assign perfectly a value to the consequence of choosing a particular alternative that it does consider [126]. If this hypothesis is valid, the rationality of the decision maker cannot be perfect [214] and it exhibits bounded rationality. March and Simon [126], suggested that the bounded rational decision maker seeks to find an alternative which meets minimally satisfactory criteria. They have hypothesized that the decision making process of the satisfying decision maker is a two stage process of discovery and selection. The first stage is that of determining the situation of the environment while the second addresses the question of what action to take in a particular situation. The stages are coupled by the dependence of the type of alternatives sought on the perceived situations.

Wohl [261] has suggested a similar two stage model of the decision process through an extension of the classical stimulus response model of psychology. When a stimulus is received, the initial reaction of the decision maker is to hypothesize about its origin. This is followed by the generation and evaluation of options, among which one response is selected.

Representations of bounded rationality have been made by Drenick, Levis, Froyd and Bailey and Papadopoulos [54, 55, 63, 164]. A characteristic of these representations is that they are based on input-output models of the decision maker. Drenick [54, 55], introduced a model where input symbols (observations) arrive and are processed into output symbols (decision responses). Each particular association of input symbol to output symbol required a fixed amount of processing time, and the bounded rationality of the decision model was represented by a maximum mean processing time.

Boettcher and Levis [32–34] introduced a basic two stage decision maker to characterize the process of executing a well defined decision making task. The fundamental departure from other models of bounded rationality is the characterization of the internal processing which is accomplished to determine output from input. Such a characterization of the decision making process is achieved by synthesizing qualitative notions of decision making with the analytic framework of Information Theory.

According to their approach, external situations are first assessed in the decision maker and then responses are selected. An information theoretic framework is used, where again the total internal activity is described in terms of internal coordination and internal decision making, as well as throughput and blockage. A constraint on the rate of total activity is suggested as a model of bounded rationality. The decision maker receives an input from its environment and uses it in the situation assessment (SA) stage to hypothesize about its origin. This results in the selection of a particular action. Possible alternatives of actions are then evaluated in theresponse selection (RS) stage. Each stage of the decision maker contains well defined procedures or algorithms. An algorithm belonging to the situation assessment stage maps input stimuli into assessed situations and an algorithm belonging to the second stage, maps assessed situations into output decision responses. Boettcher and Levis' model was restricted to decision making tasks performed in the steady state (well trained decision maker). A memoryless source was assumed where each symbol was generated independently. No learning was considered during the performance of a sequence of tasks and all mappings were deterministic.

An extension of the same work [33], includes basic interactions in an organizational context: direct and indirect control. The former is modeled as a restriction on internal decision making by external commands and the latter is incorporated through an auxiliary situation assessment input received from the organization.

Closely related to the work of Boettcher and Levis, is the work of Hall and Levis [79], in which the functions of storage and memory as they occur in a rate constrained decision making system are modeled in an information theoretic framework. They investigate two types of storage, buffer storage and memory (permanent and temporary). Buffer storage is used in processing statistically dependent inputs simultaneously and is modeled according to the input the system receives. Permanent memory is modeled as information which may be accessed by the algorithms but cannot be revised by them. Temporary memory is modeled as information that the algorithms store for use in future iterations. It is shown that Conant's partition law of information rates holds in all three cases. This approach has been tested in decision making systems in the steady state, while the same requirements and restrictions as with Boettcher's and Levis' work, hold here too.

Sen and Drenick [205], also recognized the need for adding memory to models of decision making systems. They modeled the human decision maker as an adaptive channel, a channel whose output may depend on present and past inputs. With this addition of memory they achieved results which in some experimental situations, reflected observed behavior. However, they have made no attempt to model explicitly the various types of memory that may be found in a decision making system.

Stabile and Levis [218, 219], have also studied the problem of designing the allocation of information processing tasks to organization members who interact with the organization's environment. They consider two information strategies for reducing the load on each member while accomplishing the overall task and they give special consideration to the time required for a decision maker to process information. They introduce an information theoretic interpretation of the processing time and they show that complex information structures can be constructed using combinations of parallel and alternative processing.

More recent work by Levis has extended the original two stage model to a four stage model of situation assessment, information fusion (IF), command interpretation (CI) and response selection. All previous approaches have been extended and different classes of Petri nets have been derived and utilized to reflect accurately the real time performance of the modeled systems [12].

Broekstra [38], examines the applicability of Information Theory to the problem of structure identification. He shows that structure candidates are conveniently represented by means of the concept of total constraint, which is partitioned in terms of constraints existing within elements constituting the structure candidate. For the identification of structure candidates, he uses an information theoretic measure of conformation. His work draws upon ideas developed by Klir [106].

Koomen [109], views the process of designing a system in terms of the information flow that occurs during the design. He shows how information is intimately linked with the structural complexity of the system and considers the design process as a sequence of models each representing a description of the relevant system quantities at a certain level of detail. In order to improve the model, he introduces the entropy of the design measure depending on the number of components in a model and the ways these are interconnected.

Jaynes [93, 94], shows that Information Theory provides a constructive criterion for setting up worst case probability distributions on the basis of partial knowledge, and leads to a type of statistical inference which is called the *maximum entropy* estimate. Jaynes shows that it is the least biased estimate possible on the given information, i.e., it is maximally non committal with regard to missing information. Similar work is that of Van Campenhout and Cover [244].

As far as the application of Information Theory to modeling general feedback control and communication systems, effective use of the entropy (or uncertainty) function was first demonstrated by Shannon who proved it to be a very powerful tool for evaluating the performance of communication systems. Since then, very little work has been accomplished in extending the original concepts to other applications. However, there was a strong suspicion (due to Price and Viterbi) that the entropy representation of a random process was in some way related to mean square estimation theory [171, 247]. Goblick in [74] was the first one who made a substantial contribution toward embedding the estimation problem in the framework of communication theory. His work initiated a strong motivation to suspect that the entropy function may be usefully applied to the class of systems that may be analyzed by the use of mean square error techniques. As a result, components of estimation, feedback control and adaptive control systems, being members of such a class, should be described as information transformations. Weidemann [256], analytically explains the above ideas and shows that the application of entropy techniques to the feedback control problem led to two important results. The first is a separation theorem and the second an information theoretic interpretation of the feedback process.

Kalata and Priemer in a series of four papers [96–99], use information theoretic concepts to:

1. Obtain conditions for a minimax error entropy stochastic approximation algorithm to estimate the state of a nonlinear discrete time system based on noisy linear measurements of the state.
2. Analyze the smoothing problem for linear discrete time systems using a smoothing rate upper bound formula.
3. Derive a procedure for identifying a parameter of a stochastic linear discrete time dynamic scalar system based on noisy linear measurements of the system state, and,
4. Derive fundamental principles for the general estimation problem.

In the third case after simplifying approximations the desired error entropy identification algorithm reduces to an on line adaptive identification algorithm, similar to well established identification techniques. In the fourth case the derived minimal error entropy estimator for linear systems disturbed by Gaussian random processes, turned out to be identical to a Kalman filter.

Saridis [191, 195], has shown that the equivalence of the average Lagrangian to an entropy established in Theoretical Thermodynamics may be used to express the average value of the performance criterion of a feedback control as an entropy. The optimal control problem may thus be recast as one which minimizes the entropy of selecting the feedback controls. Details about this approach are given in the next section.

2.3 The Duality of the Concept of Entropy and its Application to Control Theory

The higher levels of hierarchically intelligent systems process various types of information. Learning, knowledge acquisition, planning, decision making and data storage and retrieval, are characteristic functions of the intelligent part of such systems. These functions are directly related to information processing; so it is natural to assume that they contribute to the production of entropy in the information theoretic sense of Shannon [207, 208]. Therefore, entropy is an analytic measure of the intelligent function of the system and its rate may describe the amount of knowledge (information) processed by the general systems theoretic approach of Conant [43–46].

Considering entropy as an energy function, defined by the second law of thermodynamics and used by Boltzmann in the Theory of Statistical Thermodynamics to serve as a measure of the uncertainty of particle motion in a perfect gas as described in Lindsay and Margenau [122], proves that this function is a compatible measure for the precision of hardware execution of the tasks performed at the lowest level of the hierarchy, as it was with the higher levels.

In order to bring into perspective the dual meaning of the concept of entropy, define Shannon's (negative) entropy as a measure of the uncertainty of transmission of information:

$$H = -\int_{\Omega_S} p(s)\log p(s)ds \qquad (2.3.15)$$

Ω_s is the space of information signals transmitted, while the base of the logarithm may be defined accordingly.

Information Theory borrowed the concept of entropy from the Theory of Statistical Thermodynamics, where Boltzmann's theory states that the entropy S of a gas changing states isothermally at temperature

T is given by [36]:

$$S = k \int ((y - E)/kT) exp((y - E)/kT) dx \qquad (2.3.16)$$

where y is Gibbs energy given by:

$$y = -\int kT \log(exp(-E/KT)) dx \qquad (2.3.17)$$

E is the total energy of the system and k is Boltzmann's constant. Due to the uncertainties involved in describing the dynamic behavior of such systems, a probabilistic approach is (usually) followed. In this case, the entropy is a measure of the uncertainty of molecular distribution. Assuming that $p(x)$ is the probability of a molecule being in state x, then:

$$p(x) = exp((y - E)/kT) \qquad (2.3.18)$$

The function $p(x)$ is constrained to satisfy the incompressibility property in the state space Ω_x:

$$dp/dx = 0 \qquad (2.3.19)$$

This property of $p(x)$ is a differential constraint, which yields for perfect gases Liouville's equation [172]. Substitution of (2.3.17) into (2.3.16) gives:

$$S = -\int_{\Omega_x} p(x) \log p(x) dx \qquad (2.3.20)$$

This equation, is equivalent to (2.3.15) used to describe the entropy of an information source when the state space of the molecules of a perfect gas Ω_x is replaced by the space of information signals Ω_s. In both cases, the entropy is a measure of uncertainty about the state of the system, obtained as a function of a probability density exponential function of the involved system energy. If the dynamical theory of thermodynamics is applied on the average on all the molecules of the gas, an average Lagrangian $\bar{L}$ is defined to describe the average phenomena over time of the state x of a perfect gas, each molecule of which has a Lagrangian (the difference between the kinetic and potential energy, respectively):

$$L(x,t) = P_1 - P_2 \tag{2.3.21}$$

The average Lagrangian is:

$$\bar{L} = \int_0^t L(x,\tau)d\tau \tag{2.3.22}$$

When the average Langrangian is minimized, it satisfies the second law of thermodynamics. Since (2.3.15) and (2.3.20) correspond to the same phenomenon, it may be identified that for a perfect gas changing states isothermally, the following relation is true:

$$S = \bar{L}/T \tag{2.3.23}$$

T is the constant temperature of the isothermal process. This is the key relation to express the performance measure of a control system as an entropy.

It is known that Optimal Control Theory [19], has utilized a non negative functional of the states of the system $x(t)\epsilon\Omega_x$, where Ω_x is the state space, and a specific control $u(x,t)\ \epsilon\Omega_u$ where Ω_u is the set of all admissible controls, $\Omega_u \subset \Omega_x$, to define the performance measure for some initial conditions x_o, t_o representing a generalized energy function, the average value of Lagrangian $L(x,u,t)$ of the system. Its form is:

$$V(x_0,u,t_o) = \int_{t_o}^t L(x,u(x,\tau),\tau)d\tau \tag{2.3.24}$$

where $L > 0$, subject to the differential constraints dictated by the underlying process:

$$\dot{x}(t) = f(x, u(x,t), t),\ x(t_0) = x_s;\ x(t_f)\epsilon M_f \qquad (2.3.25)$$

with M a manifold in Ω_x. At $u^*(x,t)\ \epsilon\Omega_u$:

$$V(x_0, u^*, t_o) = \min_u \int_{t_o}^{t} L(x, u, \tau) d\tau \qquad (2.3.26)$$

Saridis [195], has shown that by selecting the density of the uncertainty of design over the space of admissible controls to satisfy Jaynes principle of maximum entropy, the associated entropy is:

$$H(x_0, u, p(u)) = E\{V(x_0, u, t)\} \qquad (2.3.27)$$

Then, the optimal control u^* satisfies:

$$H(x_0, u^*, t) = \min_u H = \min_u \{E\{V(x_0, u, t_o)\}\} \qquad (2.3.28)$$

Thus, the problem is recast as the optimal design under the worst uncertainty of selecting the best control from the space of admissible controls Ω_u covered by a density function $p(x_0, u)$ which attains its maximum value at u^*. Minimization implies:

$$dH/du = \partial H/\partial u + (dH/dp)(\partial p/\partial u) = 0 \qquad (2.3.29)$$

and since the probability density function $p(x_0, u)$ maximizes the (differential) entropy H according to Jayne's principle i.e.:

$$dH/dp = 0 \qquad (2.3.30)$$

then:

$$dH/du = \partial H/\partial u \qquad (2.3.31)$$

Based on (2.3.27), the following Theorem was proven [195]:

THEOREM: *A necessary and sufficient condition for u*(x,t) to minimize $V(u(x,t), x_o, t_o)$ subject to (2.3.25), is that u* minimizes the differential entropy H, where $p(x_0, u)$ is the maximum entropy density function according to Jaynes' maximum entropy principle.*

Entropy satisfies the additive property and any system composed of a combination of subsystems will be optimal by minimizing its total entropy.

2.4 Hierarchically Intelligent Systems

The design of complex hierarchically intelligent systems involves utilization and implementation of concepts and ideas from the diverse disciplines of Science, Mathematics and Engineering [1, 3–10, 12–16, 32–34, 40, 43–46, 81, 82, 184–202, 231–241]. Operation of such systems requires distributed problem solving techniques. Two modeling approaches are currently being followed: distributed Artificial Intelligence with emphasis to the development of actual distributed problem solving systems (multi agent planning, negotiation, functionally accurate, cooperative), and theoretical computer science and control theory with emphasis to the specification of formal processes. Wang [249], gives a thorough literature review on the subject.

From the Artificial Intelligence point of view, central issues in plan generation are the state representation, the system architecture and the planning strategies [58, 78, 120, 174, 175, 221, 259]. Planning strategies include depth or breadth first search, A* algorithm and means-ends analysis [157, 168, 170]. Artificial Intelligence planning systems utilize explicit symbolic, logical or temporal logic models of discrete operations to construct sequences of tasks to achieve specific task goals either at the motion and path planning level (using purely geometric search, potential function mapping algorithms, distance metric fields), or at the task planning level (which is beyond the kinematic and dynamic description of robotic systems) [16, 128–130]. In Artificial Intelligence expert systems, the problem domain is considered to be the knowledge base and the planner the inference engine [16, 128].

Artificial Intelligence planning approaches have dominated most of the research in robot assembly task planning using domain independent methods. As explained in [181], the central idea of these methods is to have one general purpose inference engine that may be used for any domain by describing initial state, goal and the operations in a logic formalism. Systematic procedures for task planning and representation are needed to guarantee that every assembly sequence has been generated. Most assembly type planners are hierarchical planners that use data structures such as graphs or trees to represent objects [50–53, 58, 73, 101, 181, 182]. A thorough review and analysis of different types of planners may be found in [146–148]. Moreover, the use of an automaton representation of the robot task has been suggested in [73] since it provides a natural way of mapping events to actions. Artificial Intelligence planning problems in a Petri net framework have also been investigated [14, 15].

From the system theory point of view, efforts to model intelligent systems operating autonomously or semi autonomously in uncertain environments include Saridis' work who has proposed a tree like structure defined as Hierarchically Intelligent Control Theory [184, 186, 196] and Meystel's nested hierarchical information structure [138–144] which although not tree like is beneficial and applicable to problems formulated within the theory of intelligent controls. A third approach has been followed by Albus, who has also proposed a theory of intelligence for biological and artificial systems [3, 6, 8]. An approach emphasizing the autonomicity concept in intelligent systems has been developed by Passino and Antsaklis [13–16, 167] for autonomous control systems designed to perform well under uncertainties in the system and environment for long periods of time and compensated for system failures without external intervention. Acar and Ozguner have adopted an axiomatic approach to derive a unified hierarchical control strategy that utilizes both mathematical and functional system representations [1]. A reliability analysis of Intelligent Machines is provided in [131] by using concepts from Reliability and Information Theory that illustrates the principle of Increasing Precision with Decreasing Intelligence at the execution level of Intelligent Machines. The issue of structural formulation for plan generation in Intelligent Machines is investigated in [197]. Using a relational database to represent the knowledge base and a condition event net to model the planning process, it is shown that the generated plans are Petri net languages and that any planning strategy can be achieved by a supervisory planner. Techniques for modeling intelligent systems with Petri nets may be found in [14,

15, 240, 249–254], while applications of expert systems to control, named expert control, may be found in [18]. Most of the above issues are further addressed in [142]. Performance of this type of systems is improved through self-modification [185] with learning algorithms [65, 66].

Concentrating in robotic assemblies (RAs) modeled based on intelligent control methodologies, the important issue related to the derivation of automated planning techniques is the development of efficient general purpose procedures that can find at least one plan in a wide variety of situations, rather than procedures that eventually find the best plan in a more restricted type of situation [51]. Searching for the most efficient plan requires criteria to decide whether a plan is better than another. The information necessary to make this decision is usually available during the execution of the plan. Hence, there is a choice between off line and on line planning. Task planning is critically dependent on the task representation. Systematic procedures are needed to guarantee that every assembly sequence has been generated. Several representation schemes have been proposed to store all possible assembly plans for a given product that include ordered lists of actions, triangle tables, precedence diagrams and AND/OR graph representations [51].

Ordered lists of actions do not permit flexibility in executing a task. Triangle tables improved the capability to recover from errors but only within one fixed sequence [51, 58]. Fox and Kempf addressed the need to act opportunistically as opposed to always following a preprogrammed fixed order of operations. They suggested that plans generated off line be given to the robot as a set of operations with minimal ordering constraints. A plan is represented by a precedence diagram and actually encompasses several possible sequences of operations that perform the task of assembling a given product [101]. The problem with the precedence diagram formalism is that for most products no single partial order can encompass every possible assembly sequence. The AND/OR graph representation suggested by de Mello and Sanderson [51–53, 59] encompasses all the possible ways to assemble a product, and therefore allows one to explore the space of all possible plans. Since plans correspond to solution trees in an AND/OR graph, the solution of the best plan can be seen as a search problem. By assigning cost functions to the hyperarcs (that correspond to the disassembly operations), the best plan corresponds to the solution tree that has the minimum cost. Smith and Gini [73] suggested the use of an automaton representation of the robot task since it provides a natural way of mapping events to actions. Significant sensor readings are treated as 'tokens' that may cause state transitions and actions related thereto. The discrete states in the automaton representation also provide a useful way of identifying separate parts of the robot's task for automated reasoning and analysis. Other Artificial Intelligence techniques used to describe and represent planning and problem solving capabilities of robotic systems suffer from being computationally expensive in memory and time.

Another important class of complex systems benefited from intelligent control methodologies are manufacturing systems. Kusiak [110, 111] describes significant developments in manufacturing technology. Artificial Intelligence as applied to manufacturing systems is the theme of several review papers [61, 102]. The resource allocation problem in distributed factory scheduling is described in [176]. A hierarchical approach for modeling Flexible Manufacturing Systems with Petri nets is presented in [240]. An object oriented environment for robot system architectures has appeared in [119] where the system architecture is based on hierarchies of distributed multiprocessors forming a layered programming structure.

In addition, two formal hierarchies have already been proposed by the Computer Aided Manufacturing International (CAM-I) and National Institute of Standards and Technology (NIST) (former NBS). The CAM-I structure has four levels [47] and the NIST (former NBS) system architecture has five [9]. However, O'Grady has shown that [162, 163] "... *both hierarchies have a rough equivalence in function and levels. The extra layer in the NIST hierarchy mainly results from the use of virtual manufacturing cells."* Both of these structures are fundamentally different from the proposed structure in this monograph in that the number of levels is not fixed.

An intelligent cell control system for automated manufacturing, a knowledge based approach to industrial job floor scheduling, and intelligent information models for operating automatic storage and retrieval systems are presented in [111]. Hierarchical approaches to control manufacturing processes have been suggested in [62, 162, 163, 209].

However, as explained in [183], the major obstacle in designing complex systems is the *overall system coordination* and not the control of specific devices. Other major obstacles include implementation programming and planning of various devices in order to create a reliable system. The main objective of such systems is to synthesize and perform a sequence of actions so that the overall system goal is achieved even when replanning is necessary.

2.5 Knowledge Based Expert Systems

Artificial Intelligence and its subfield knowledge based expert systems are equipped with the capability to represent knowledge in different forms. Practical results may be attributed to the design and use of knowledge based expert systems, problem solving computer programs that can reach a level of performance comparable to that of a human expert in some specialized problem domains like, Computer Systems, Computing, Education, Engineering, Knowledge Engineering, Fault Diagnosis, Geology, Medicine and Science.

What distinguishes an expert system from an ordinary applications program is the model of problem solving [81, 82, 105]. Applications programs make use of specialized problem solving knowledge and many of them reach high levels of performance. The model of problem solving in expert systems is based on a separate entity or knowledge base, while in an application program it appears implicitly, as part of the coding of the program. As a result, expert systems instead of being programmed to follow step-by-step procedures, follow a few general procedures for finding solutions to problems. Facts, heuristics, models and other general knowledge about solving a particular class of problems are encoded and stored in the computer's memory. In order to solve a particular problem, the computer uses facts by the user plus its domain knowledge and general problem solving procedures to find and apply a specific solution. This knowledge is acquired from a domain expert who provides the key to expert performance. Knowledge representation and inference schemes provide the mechanisms for its use. Encoding of knowledge in program form is accomplished by the knowledge engineer, who proceeds through several stages before producing the knowledge based system.

A typical knowledge based expert system consists of an inference engine, a knowledge base and a workspace [40], as shown in Figure 2.1. The inference engine solves a problem by interpreting the domain knowledge contained in the knowledge base. The workspace is an area of memory set aside for storing a description of the problem constructed by the system from facts supplied by the user, or inferred from the knowledge base during a consultation. It may also include a natural language interface for communicating with a user, a reasoning explanation subsystem and a knowledge acquisition subsystem for expanding the knowledge base. The main point of interest is the explicit division between the inference engine and the knowledge base. Two important advantages result from keeping this division as strict as possible. First, if all of the control structure information [60, 105] has been kept in the inference engine, then one can engage the domain expert in a discussion of the knowledge base alone rather than of questions of programming and control structures. Second, if all of the task specific knowledge has been kept in the knowledge base then it is possible to remove the current knowledge base, "plug in" another and obtain a performance program for a new task. This does not mean of course that the inference engine and the knowledge base are

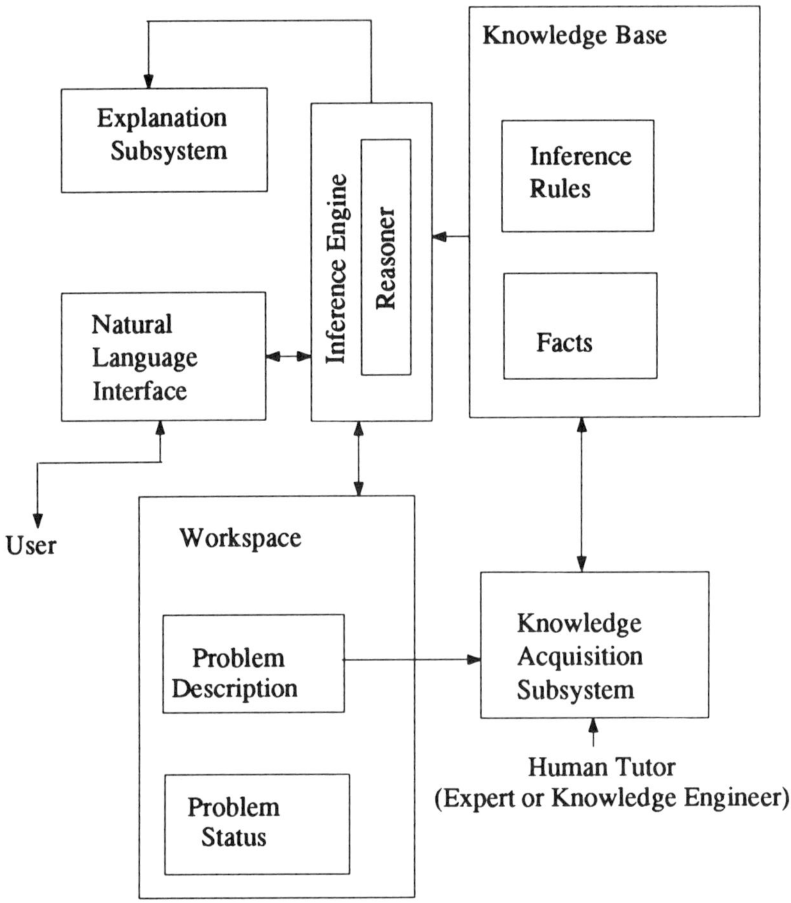

Figure 2.1 A Typical Expert System

completely independent. The knowledge base content is influenced by the inference engine, since the rules written for the knowledge base take into account the inference engine and its built in control strategies.

Buchanan, Shortliffe and Hayes-Roth, define knowledge in a specific domain to consist of descriptions, relationships and procedures [40, 81, 82]. The descriptions are sentences in some language that characterize the definitional and empirical relationships in the domain and the procedures are manipulating these descriptions. To represent knowledge means to engineer knowledge, that is, convert it to applicable forms. An assortment of techniques is used for representing knowledge, including:

1. Production rules.
2. Semantic networks.
3. Frames, and,
4. First order logic.

Each of these schemes has advantages. Production rules are particularly useful for representing procedural knowledge methods for accomplishing goals. Semantic networks are good for representing relations among objects. Frame based semantic networks can concisely store a vast amount of knowledge about object properties and relations. First order logic supplies the means for explicitly expressing virtually any type of knowledge. All the above representations are computationally equivalent to the Turing machine. Although they produce very sophisticated programs capable of solving tasks which require extensive human expertise, they have an extremely narrow range of expertise and they are unable to recognize when a given problem is outside this range of expertise. They cannot solve simpler versions of the problems they are designed to solve [37].

2.6 Comparison Between Knowledge Based Expert Systems and Intelligent Machines

We conclude this chapter with a brief comparison between knowledge based expert systems and Intelligent Machines as discussed in this monograph.

It is postulated that the capabilities and limitations of knowledge based expert systems and Intelligent Machines are related to their knowledge about the specific problems they are faced with and their ability to apply this knowledge to solve such problems. This knowledge is represented within a knowledge based expert system and/or Intelligent Machine in different but equivalent ways.

In a knowledge based system, knowledge is composed of:

1. Knowledge representation.
2. Reasoning.
3. Natural Languages, and,
4. Cognition.

Since the highest level of Intelligent Machines is modeled after a knowledge based system (as it will become evident in the next two chapters), in this level the previous components respectively correspond to:

1. Data handling and management.
2. Inference.
3. Formal languages, and,
4. Sensing.

However, both knowledge based systems and Intelligent Machines utilize the power of the digital computer to operate efficiently and solve the problems they have been designed for. The corresponding components of the digital computer to the above four functions are:

1. Memory.
2. CPU.
3. Operating system, and,
4. Input-Output Devices.

But although there is a fundamental similarity in the way the digital computer is utilized to represent knowledge in knowledge based systems and Intelligent Machines, they do have two major implementation differences:

1. Even though digital computers may be programmed with human logic and heuristics, the design procedures of knowledge based systems do not consider the interaction between the different subsystems involved in such a design, and,
2. Knowledge based systems are completely interactive, i.e., they require the presence of a human expert for intelligent performance. Intelligent Machines may function non interactively because they are equipped with compatible analytic measures.

Intelligent Machines have an advantage over knowledge based systems which results from their hierarchical structure. The organizer uses hierarchical planning techniques for the formulation of the sequence of operations which will be executed at the lowest level. This hierarchical planning helps avoid exponential growth of detail that would result from planning at the most primitive level.

Knowledge based expert systems follow symbolic reasoning methods to solve the problems they are faced with. This expert reasoning is often non monotonic, requiring the making and retraction of assumptions. Intelligent Machines follow formal mathematical reasoning.

A major disadvantage common to both types of systems is their lack of common sense knowledge and their inability to solve simpler versions of the problems they have been designed for. Artificial Neural Networks are currently being used to overcome this limitation.

2.7 Remarks

A thorough review on hierarchical and expert systems has been presented in this chapter. The concept of entropy and its utilization in formulating and modeling general systems has been illustrated and justified.

The next chapter is devoted to the development of the structure of Intelligent Machines. The principle of Increasing Precision with Decreasing Intelligence is presented as an analog to the theory of organizations and it is proven for the rate of knowledge, the structured information as the main variable of the system. Then, the three level structure of the machine is explained and newer architectures are suggested. Finally, the applications of the theory to Intelligent Robotic Systems, a special implementation of Intelligent Machines, is briefly discussed.

Chapter 3
ON THE GENERAL THEORY OF INTELLIGENT MACHINES

3.1 Introduction

In the past fifteen years, several researchers have developed viable theories for Intelligent Machines and created working models to implement such theories [3–10, 137, 138–144, 146–148, 184–202, 231–241, 249–254]. Intelligent Machines were designed to perform anthropomorphic tasks with minimum interaction with a human, with potential applications to robotic systems operating in remote, inaccessible, hazardous, unfamiliar or other environments as need appeared.

Since the task was enormous and the available technologies rather limited, the results of such an effort have been meager. The theoretic efforts that have come into the picture to reinforce the development of Intelligent Machines have taken two distinct directions: the logic based approach as described by Nielson [156–159] and the analytic approach of Saridis [184–202] and Meystel [138–144].

The results on the analytic approach have been reported regularly by the authors and their colleagues [146–148, 184–202, 231–241, 249–254] and have reached a level of maturity both theoretically and experimentally. Detailed mathematical and architectural models based on a probabilistic approach and an expert system cell are given in the next chapters.

Quite recently, new methodologies and new techniques like Neural nets, Petri nets and Boltzmann machines, have appeared in the literature and have provided new tools for the analytic formulation of the theory of Intelligent Machines. The advances, development and adaptation of such tools, along with the better understanding of the process, led to further modifications and refinements of the theory aimed to strengthen, simplify and integrate the proposed design of Intelligent Machines. These new methodologies will be summarized in the last chapter.

The developments of the theory introduced herein, provide a better understanding of some of the basic concepts related to Intelligent Machines, like the principle of Increasing Precision and Decreasing Intelligence, machine knowledge, machine intelligence and the flow of knowledge within such machines, and set the fundamentals necessary to understand and develop Boltzmann machines and Petri nets as realizations of inference engines and linguistic decision schemata, respectively. These recent modeling approaches, still utilize entropy measures for the evaluation of the performance at every level of the machine.

The analytic formulation of intelligent controls is presented in the next section, followed by a set of definitions and a discussion on the principle of Increasing Precision with Decreasing Intelligence. Details on the three interactive levels of Intelligent Machines are then discussed. Intelligent Machines are also placed in a distributed intelligence environment, followed by applications, discussions and conclusions. The fundamentals described in this chapter, apply directly to all types of hierarchical systems, including Intelligent Robotic Systems.

3.2 Review of the Analytic Formulation of Intelligent Controls

Intelligent Machines require the use of generalized control efforts in order to perform intelligent functions such as simultaneous utilization of a memory, learning, or multi level decision making in response to simple qualitative commands.

Intelligent controls have been developed to implement such functions in hierarchical systems and Intelligent Machines. Intelligent controls may be considered as a *fusion* between the mathematical and linguistic methods and algorithms applied to systems and processes. Both methods utilize the results of cognitive systems research effectively with various mathematical programming control techniques. However, they lack that rigor and precision that pure analytic techniques provide.

The control intelligence in Intelligent Machines is hierarchically distributed according to the principle of Increasing Precision with Decreasing Intelligence, evident in all hierarchical management systems, which is further discussed and proven in the sequel. The resulting structure is composed of three basic levels of controls, each level of which may contain more than one layer of tree structured functions [184–202, 231–241]. The three interactive levels of organization, coordination and execution are shown in Figure 3.1.

The functions involved in the upper levels of the machine are imitating functions of human behavior and may be treated as elements of knowledge based systems. The functions of planning, decision making, learning, data storage and retrieval and task coordination, may be thought of as knowledge handling and management. On the other hand, the system control problem has been shown to be reformulated by using entropy as a control measure and therefore integrate all the hardware activities associated with the machine with the higher levels, regardless of the discipline they belong to. Thus,

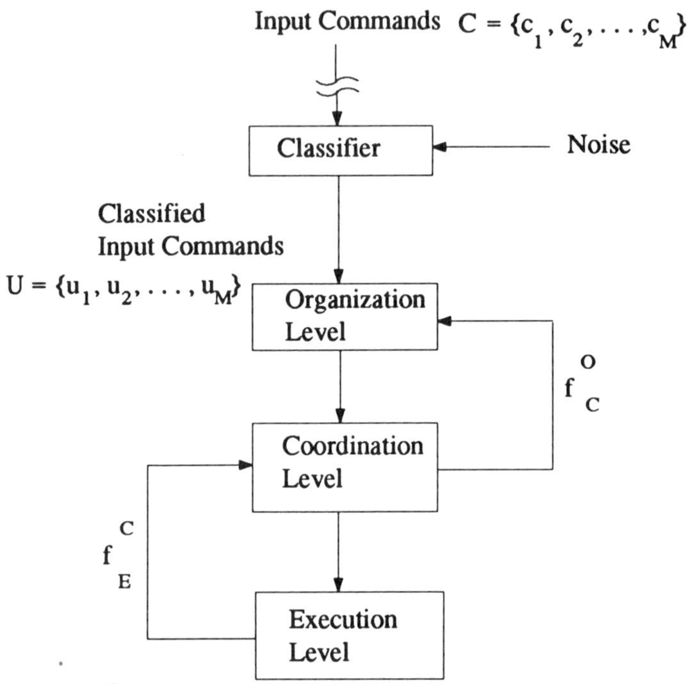

f_E^C : Feedback from the execution to the coordination level (on-line)

f_C^O : Feedback from the coordination to the organization level (off-line)

Figure 3.1 Structure of an Intelligent Machine

vision coordination, motion control, path planning and force sensing in a robot paradigm, may be integrated into the pertinent actions of the machine and evaluated by common entropy functions. Therefore, *the flow of knowledge may be considered as the key variable* of such a system.

Knowledge flow in an Intelligent Machine represents respectively:

1. Data handling and management.
2. Planning and decision performed by the central processing units.
3. Sensing and data acquisition obtained through peripheral devices, and,
4. Formal languages which define the software.

Since all levels of a hierarchically intelligent control system can be measured by entropies and their rates, then the optimal operation of an Intelligent Machine may be obtained through the solution of mathematical programming problems, summarized as follows: *The Theory of Intelligent Controls may be postulated as the mathematical problem of finding the right sequence of decisions and controls for a system structured according to the principle of Increasing Precision with Decreasing Intelligence (constraint) such that it minimizes its total entropy.*

3.3 Definitions and the Principle of Increasing Precision with Decreasing Intelligence

It remains to investigate the general concepts of intelligent control systems which pertain to the fundamental functions of Intelligent Machines. In order to establish an information theoretic framework for the system model, it is necessary to clarify the actual meaning of the terms knowledge, intelligence and information, and then, give appropriate definitions for machine knowledge, machine intelligence, machine precision and machine imprecision.

Quoting Webster [255], "*....knowledge is the fact or condition of knowing something with familiarity gained through experience or asso-*

ciation; the range of one's information or understanding. Information is the communication or perception of knowledge; knowledge obtained from investigation, study or instruction; a quantitative measure of the content of information. Intelligence is the ability to apply knowledge to manipulate one's environment or to think abstractly as measured by objective criteria". Therefore, it should be clear that the term "knowledge" is more general than the term "information". Information Theory may be utilized for the mathematical interpretation of machine knowledge and intelligence because it is *"...the theory that deals statistically with information, the measurement of its content in terms of its distinguishing essential characteristics, and with the efficiency of processes of communication between men and machines"*.

The corresponding terms for the design of Intelligent Machines are defined as:

Definition 1. ***Machine Knowledge (K) is defined to be the structured information acquired and applied to remove ignorance or uncertainty about a specific task pertaining to the Intelligent Machine.***

Machine knowledge within an Intelligent Machine includes both a priori and a posteriori knowledge. A priori knowledge is acquired through any initial information given by the system designer and/or user. A posteriori knowledge is acquired and accumulated through learning and experience. Machine knowledge is represented within the Intelligent Machine *explicitly* and *implicitly*. Explicit knowledge representation requires data, data bases and knowledge bases, semantic nets, etc. Implicit knowledge representation is embedded within the system control and learning algorithms.

The term machine knowledge indicates the total amount of information that has been *accrued* in the long term memory of the organizer. It includes *memory operations* like storage and retrieval of information. Retrieval of information is needed for the internal decision making procedures of the organizer, while storage of information, performed through feedback upon completion of the requested job, is necessary to upgrade previously stored information.

Definition 2. ***The Rate of Machine Knowledge (R) is the flow of knowledge through an Intelligent Machine.***

The American Heritage Dictionary of the English Language defines intelligence as follows: *"Intelligence is the capacity to acquire and apply knowledge"*. In terms of Machine Intelligence, this definition may be modified to yield:

Definition 3. ***Machine Intelligence (MI) is defined as the process of analyzing, organizing and converting data into knowledge. Machine Intelligence, is the set of actions or rules which operates on a Data Base (DB) of events or activities to produce flow of knowledge (R).***

On the other hand, one may define machine precision and imprecision as follows:

Definition 4. ***Machine imprecision is the uncertainty of execution of the various tasks of the Intelligent Machine.***

Definition 5. ***Machine precision is the complement of machine imprecision and represents the complexity of the process.***

The above definitions may be also derived mathematically to provide an analytical interpretation of the system concepts.

Machine knowledge (K) representing a type of accrued information may be represented as:

$$K = -\alpha - lnp(K) \tag{3.3.1}$$

where $p(K)$ is the probability density of knowledge and α is an appropriately chosen constant.

From the above equation, the probability density function $p(K)$ satisfies the expression in agreement with Jaynes principle of maximum entropy [93, 94]:

$$p(K) = e^{-\alpha - K}; \; \alpha = \ln \int_{\Omega_s} e^{-K} ds \tag{3.3.2}$$

where all the terms have been defined previously. The rate of knowledge R which is the main variable within an Intelligent Machine with discrete states, is defined over a fixed interval of time T:

$$R = K/T \tag{3.3.3}$$

It was intuitively thought [196], that the rate of knowledge must satisfy the following relation characteristic of the principle of Increasing Precision with Decreasisng Intelligence:

$$(MI) \ : \ (DB) \, to \, (R) \tag{3.3.4}$$

as evidenced by Definition 3. An analytic formulation of the above principle derived from simple probabilistic relations among the rate of knowledge, machine intelligence and the data base of knowledge may be now derived and justified. The entropies of the various functions come naturally into the picture as a measure of their activities.

3.4 The Analytic Formulation of the Principle of Increasing Precision with Decreasing Intelligence

In order to formulate mathematically the concepts of knowledge based systems, one must consider the state space of knowledge Ω_s with states $s_i, i = 1, 2, ..., n$. They represent the state of events at the nodes of a network defining the stages of a task to be executed.

In an Intelligent Machine knowledge between two states is considered as the association of the state s_i with another state s_j and is expressed as:

$$K_{ij} = 1/2 w_{ij} s_i s_j \tag{3.4.5}$$

where w_{ij} are state transition coefficients, which are zero in case of inactive transmission.

Knowledge at the state s_i is the association of that state with all the other active states s_j and is expressed as:

$$K_i = 1/2 \sum_j w_{ij} s_i s_j \tag{3.4.6}$$

Finally, the total knowledge of a system is considered as:

$$K = 1/2 \sum_i \sum_j w_{ij} s_i s_j \tag{3.4.7}$$

and has the form of energy of the underlying events. The rate (flow) of knowledge is the derivative of knowledge and for the discrete state space Ω_s is defined respectively as:

$$R_{ij} = K_{ij}/T, \ R_i = K_i/T, \ R = K/T \tag{3.4.8}$$

where T is a fixed time interval.

The principle of Increasing Precision with Decreasing Intelligence is expressed probabilistically by the relation:

$$Prob(MI, DB) = Prob(R) \tag{3.4.9}$$

where MI is the machine intelligence and DB is the data base associated with the task to be executed and represents the complexity of the task which is also proportional to the precision of execution. The following relation is obtained by conditioning and taking the natural logarithms:

$$lnp(MI/DB) + lnp(DB) = lnp(R) \tag{3.4.10}$$

Taking the expected value on both sides one obtains the entropy equation:

$$H(MI/DB) + H(DB) = H(R) \tag{3.4.11}$$

where $H(x)$ is the entropy associated with x, as previously defined. For a constant rate of knowledge which is expected during the conception and execution of a task, increase of the entropy of DB requires a decrease of the entropy of MI for the particular data base, which manifests the principle. If MI is independent of the DB then:

$$H(MI) + H(DB) = H(R) \tag{3.4.12}$$

A detailed proof of the principle is given in [196].

This principle is applicable both across one level of the hierarchy as well as through the levels of the hierarchy, in which case the flow R represents the throughput of the system in an information theoretic manner. The partition law of information rates applies naturally to such a system as shown in [232]. A case study demonstrating the validity of the above is given in [131, 196].

3.5 The Analytic Structure of an Intelligent Machine

In order to implement an Intelligent Machine on analytic foundations, the theory of intelligent controls is utilized. This theory assigns analytic models to the various levels of the machine and improves them through a generalized concept of selective feedback as explained in the next chapter.

The intelligent control system is composed of three levels in decreasing order of intelligence and increasing order of precision as stipulated by the principle of Increasing Precision with Decreasing Intelligence. The number of levels is crucial and is not affected by our current better understanding of the basics and the proposed new methodologies to analytically implement the various functions of an Intelligent Machine [249–254]. Moreover, each level model is not significantly changing.

The system is viewed as an entity of transforming a (qualitative) user command to a sequence of mechanical actions. The output of the system is realized via specific commands to a set of actuators. Once the initial user command has been received, the system actions depend on input information from a set of sensors interacting with the environment. These *external* and *internal* sensors provide information to monitor the workspace environment (external) and the status of each subsystem (internal), e.g., position, velocity and acceleration in a robot paradigm. The Intelligent Machine fuses the information and chooses between alternate actions.

The organization level is designed to organize a sequence of abstract actions, general tasks and rules from a set of primitives stored in a long term memory regardless of the present world model. In other words, it serves as the generator of the rules of an inference engine by processing high levels of information for machine reasoning, planning, decision making, learning (feedback) and memory operations as shown in Figure 3.2.

The coordination level is an intermediate structure serving as an interface between the organization and execution levels. It deals with real time information of the world by generating a proper sequence of subtasks pertinent to the execution of the original command. It involves coordination of decision making and learning on a short term memory, e.g., a buffer. It utilizes linguistic decision schemata with learning capabilities [76, 77], and is assigned subjective probabilities for each action. The respective entropies may be obtained directly from these subjective probabilities. The coordination level is composed of a specific number of coordinators of fixed structure, each performing specific functions. Communication between the coordinators is accomplished through a dispatcher, the variable structure of which is dictated by the organizer, as shown in Figure 3.3.

The execution level performs the appropriate control functions on the processes involved. Their performance measure can also be expressed as an entropy, thus unifying the functions of an Intelligent Machine.

The three level structure of Figure 3.1, has the potential of *top-down* and *bottom-up* knowledge (information) processing. The top-down flow of knowledge is related to task generation and decomposition at finer levels of detail. The bottom-up flow of knowledge is related to aggregation of selective information from simple feedback at the lowest layer of the execution level (as defined in Classical Control Theory) to accumulation of knowledge at the highest level. This feedback information is essential for learning. Learning in Intelligent Machines is required and is necessary to choose from alternate actions.

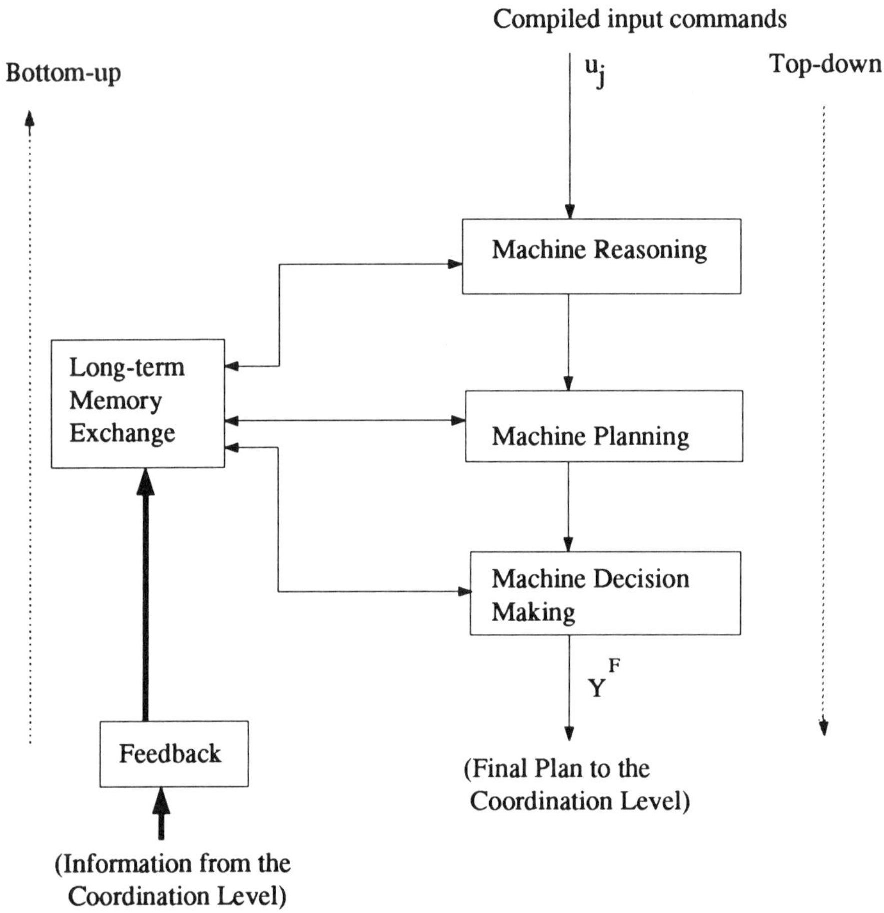

Figure 3.2 Block Diagram of the Organization Level Functions

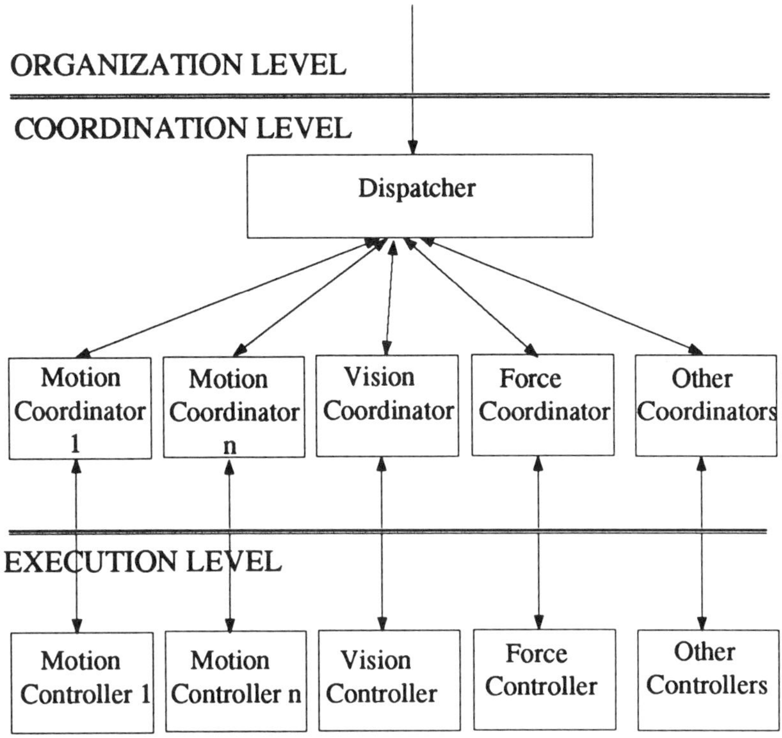

Figure 3.3 General Structure of the Coordination Level

The aggregation of information is used to update continuously the knowledge base of the system.

The principle of Increasing Precision with Decreasing Intelligence is applicable at every level of the machine, reaffirming its universal validity. However, the coordination may serve as a salient example of its application where the intelligence provided by the organization level as a set of rules is applied to the data base provided by the execution level to produce the flow of knowledge.

Optimal Control Theory, designed mainly for motion control, may be implemented for vision control, path planning and other sensory system pertinent to a machine by slightly modifying the system equations and cost functions. After all one is dealing with real time dynamic systems which may be modeled by a dynamic set of equations.

3.6 Intelligent Machines in a Distributed Intelligence Environment

In the real world, distributed systems and hierarchical systems coexist in harmony. The human organism is a typical example of this statement.

Distributed Artificial Intelligence (DAI) is a discipline concerned with treating problems that require multiple solvers in parallel by invoking Artificial Intelligence techniques [49]. When utilized to control Intelligent Machines working in parallel, it may be interpreted as Distributed Machine Intelligence (DMI) where the intelligence processing is referred to the autonomous abilities of the machines involved as with simple hierarchically intelligent control cases [195–200, 202]. This corresponds more to the distributed problem solving process and may be thought of as composed of two components related to Distributed Machine Intelligence:

1. Control, and,
2. Communications.

On the other hand, distributed control can be performed in two ways:

1. Control by a meta level, and,
2. Control by majority vote.

The first method is an extension of the hierarchical approach where the coordination, decision making and subtask assignment is deferred to a higher level of intelligence imbedded in the dispatcher of the Intelligent Machine. The cooperative activities should be planned, scheduled and sequenced in this device and communicated continuously to the machines. Feedback from the environment should be communicated continuously for the evaluation of the team work performed.

The second method deals with cooperative approach of machines operating in the same environment and performing tasks that require scheduling and task assignment. Majority vote may provide the proper planning and sequencing of the various tasks to be performed in unison by all machines involved. The majority vote could be taken in a poll place equally accessed by all machines and communicated back to them in the appropriate sequence.

The communication problem plays a paramount role in Distributed Machine Intelligence. It may be performed by a large communication network in the case of wide spacially distributed machines or by a computer bus when dealing with a tightly built system of devices. The main design considerations of such a communication system are:

1. The system configuration.
2. The protocol, and,
3. The treatment of uncertainty of information.

The first item deals with the selection of the proper structure of the network. Two types suitable for the appropriate control categories are the star connection and the ring connection.

The second item is essential for the most efficient operation of the system and the optimization of the information exchange among the machines. The computer literature contains many sources of information about protocols [112].

The third item deals with the ability of the communications system to deal with uncertain and incomplete information. The problem of reliability for accurate and precise transmission and reception of information is essential. The classical Shannon's Information Theory methods are applicable here [207, 208].

Finally, as mentioned earlier, Distributed Machine Intelligence may be applied to coordinate a number of cooperating machines or to organize a number of coordinators within the same machine. In both cases, such a structure can work in harmony with the hierarchically intelligent control structure. The reason is that the hierarchical stratification refers to the intelligence of the machine and the principle needs only to be generalized from a vertical to a horizontal deployment. In other words, the principle of Increasing Precision with Decreasing Intelligence should be assigned to all directions of flow of knowledge to represent all the trade offs between intelligence and complexity.

3.7 Applications to Robotic Systems

The Theory of Intelligent Controls has direct application to the design of Intelligent Robotic Systems. The principle of Increasing Precision with Decreasing Intelligence provides a means of structuring hierarchically the levels of the system. Since for a passive task the flow of knowledge through the system must be constant, it assigns the highest level with the highest machine intelligence and smallest complexity (size of data base), and the lowest level with the lowest machine intelligence and largest complexity. Such a structure agrees with the concept of most organizational structures encountered in human societies. Application to machine structures is straight forward.

At the present time there is a large variety of applications for Intelligent Machines. Automated material handling and assembly in an automated factory, automation inspection, sentries in a nuclear containment are some of the areas where Intelligent Machines have and will find a great use. One of the most important applications is the application of Intelligent Machines to unmanned space exploration where, because of the distance involved, autonomous anthropomorphic tasks must be executed and only general commands and reports of executions may be communicated.

Those tasks are suitable for Intelligent Robotic Systems capable of executing anthropomorphic tasks in unstructured uncertain environments. They are structured usually in a human like shape and are equipped with vision and other tactile sensors to sense the environment and execute tasks and locomotion for appropriate mobility in the unstructured environment. The controls of this type of machine are performed according to the presented general theory of Intelligent Machines.

Recent research has been focused in the application of the theory of Intelligent Machines to design robots for autonomous manipulation and locomotion in space. Satellite maintenance, construction of the space station and autonomous planet exploration vehicles are typical examples. A testbed for earth simulation of such activities in space has been built in the Center for Intelligent Robotics for Space Exploration at Rensselaer. The system configuration is shown in Figure 3.4.

3.8 Remarks

Analytic models for Intelligent Robotic Systems are developed in the next chapter. The realization of the system is obtained through the application of hierarchically intelligent control based on a better understanding of the principle of increasing precision (complexity) with decreasing intelligence, which utilizes the three level structure. The upper level is implemented through a probabilistic model capable of task planning at an abstract level. An expert system cell is also described for the organizer functions.

The coordination level composed of a dispatcher and several coordinators is implemented by linguistic decision schemata.

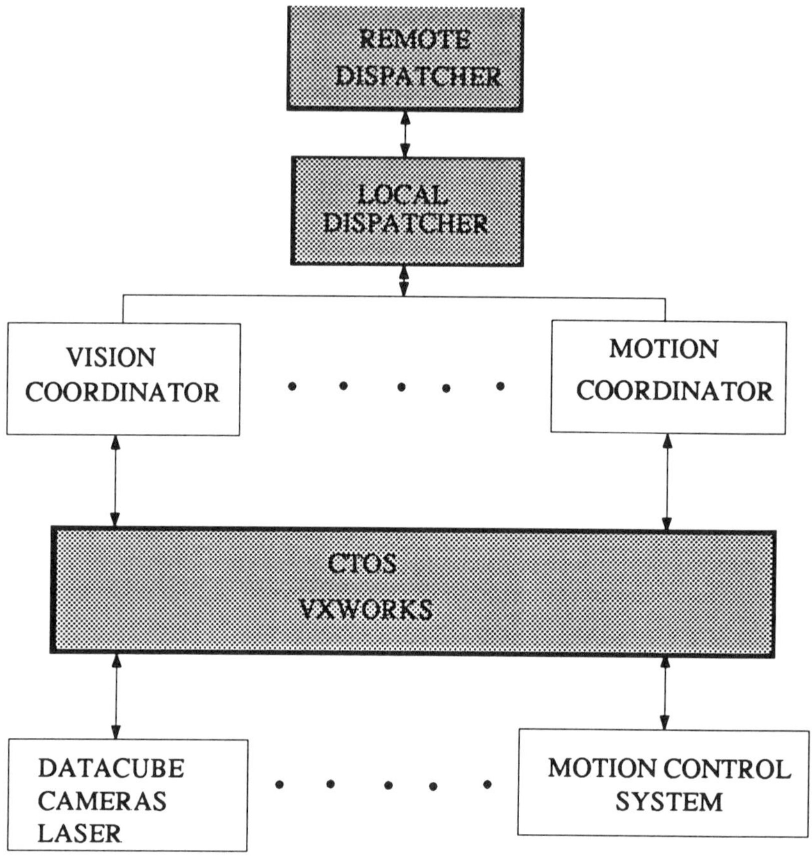

Figure 3.4 The RPI CIRSSE Telerobotics Testbed Configuration

The execution level involving motion control, visual image acquisition and sensing, is modeled by dynamic systems with entropy as a cost function, unifying the evaluation of the various processes. An integrated control and diagnostics mechanism for robot payload variations concludes the chapter.

Chapter 4
THE MATHEMATICAL MODEL FOR INTELLIGENT ROBOTIC SYSTEMS

4.1 Introduction

The purpose of this chapter is to derive the models suitable to the hierarchical structure of Intelligent Robotic Systems operating under the constraints of hierarchically intelligent control systems, define the levels individual functions and explain the flow of knowledge (information) within complex machines and systems.

The probabilistic model of the organization level is presented first. The organizer's individual functions are defined and interpreted mathematically. Operational rules imposed on the organization level functions are stated and a step-by-step mathematical algorithm follows. The expert system cell is then described. Features of the expert system cell, different from or similar to the probabilistic model are demonstrated and justified, when necessary. The organization level follows a bottom-up approach for the formulation of plans: Plans are composed from primitive events and checked for compatibility and completeness. The model of the coordination level and its functions are then defined and explained along with the coordination level algorithm. Appropriate cost functions are associated with the coordinators. The execution level is defined and interpreted in such a way to complete the hierarchical structure of Intelligent Robotic Systems. An integrated control and diagnostics mechanism capable of accommodating payload variations is described and integrated with the rest of the system. This mechanism allows the system to overcome local failures related to manipulator payload variations (e.g., accidental dropping of an object). All pertinent architectural models are presented in the next chapter.

4.2 The Probabilistic Model of the Organization Level

In order to formulate mathematically the organization level, a probabilistic model is proposed. This model uses concepts developed in knowledge based systems of Artificial Intelligence. The pertinent functions of the human brain are defined to be: reasoning, planning, decision making, feedback (learning) and long term memory exchange. They are performed sequentially in the same order within the Intelligent Robotic System. Since intelligent control is performed non interactively, it is essential to derive an efficient way of performing those functions suitable also to Machine Intelligence. A special architecture operating in machine code is thought to be the most appropriate structure that can generate fast and reliable operation of the organization level. This is obtained by defining the various operations in a mathematical way and then assigning a probabilistic structure to organize the appropriate events/actions for execution.

Although it is highly desirable to keep the general laws which govern the performance of all types of hierarchical systems independent of modeling methodologies, one should not overlook the fact that a system is designed for particular applications with more or less specific operational constraints. System goals should be met regardless of modeling approaches. Therefore, it is apparent that some 'guidelines' be given to the system during its initialization phase. This a priori (initial) knowledge (information), upon which the system will build its knowledge base and self-organization procedures, includes information related to availability of resources, tools, materials, timing requirements, etc. All this accrued information may be thought of as knowledge contributing to task completion. This a priori knowledge corresponds to "learning by a teacher" which, as the system continues to function, becomes "learning by experience" or "learning without a teacher."

This necessary a priori information is provided in terms of:

1. A set of definitions establishing the organizer workspace, and,
2. A set of operational procedures governing the organizer functions.

Both sets are general and valid for the types of systems discussed in this monograph. They may be used as guidelines for plans organization and formulation for all types of hierarchical systems. They are assigned specific meaning once the application problem is known, as demonstrated through the examples. The sets of pertinent definitions and procedures are presented in the sequel, while all the details are provided with the derived algorithm.

Pertinent Definitions in the Organizer Workspace

For every system with the proposed three level structure the following are defined:

1. The set of typed in user commands:

$$C = \{c_1, c_2, ..., c_M;\ M \text{ fixed and finite}\} \qquad (4.2.1.1)$$

with a priori probabilities $p(c_n)$, $n = 1, 2, ..., M$, sent to the system via any remote or not communication channel.

2. The set of classified compiled input commands:

$$U = \{u_1, u_2, ..., u_M;\ M \text{ fixed and finite}\} \qquad (4.2.1.2)$$

with associated probabilities $p(u_j/c_n)$, $j, n = 1, 2, ..., M$, which are the actual inputs to the organization level of the system. The classification is performed via well known classification techniques as illustrated in the sequel.

3. The task domain of the system is defined with the set:

$$E_t = \{e_1, e_2, ..., e_N;\ N \text{ fixed and finite}\} \qquad (4.2.1.3)$$

of primitive events (actions). These primitives are basically divided into the *disjoint* subsets of non repetitive and repetitive events, respectively defined as E_{nr} and E_r. Non repetitive events are considered those occurring only once within a plan (e.g. grasp, dump), while repetitive those occurring more than once (e.g. move) within a plan. Without loss of generality, consider $(N - L)$ non repetitive events and L repetitive ones within E_t. Given the task domain and the specific application problem, all subsets shown in Table 1 must be defined.

4. Define the binary valued random variable $x_i, i = 1, 2, ...N$, associated with each e_i indicating if e_i is active $(x_i = 1)$ or inactive $(x_i = 0)$ within a plan given a specific u_j, with corresponding probabilities $p(x_i = 1/u_j) = p_{ij}$ and $p(x_i = 0/u_j) = 1 - p_{ij}$, respectively.

5. Define the set of activities A_{jm} (groups of primitive events concatenated together to form a complex task) associated with a specific u_j, and are represented in terms of the binary string X_{jm} indicating which events/actions are active and/or inactive within an activity A_{jm}. The indices indicate the mth activity-string associated with the jth compiled command. Since the x_i's are considered binary, the initial maximum number of activities A_{jm} (strings X_{jm}) associated with one u_j is $(2^N - 1)$. The corresponding

probability of an activity-string is defined as:

$$P(X_{jm}/u_j) = \left\{ \prod_{i=1}^{N} p(x_i/u_j) \right\}_m \quad : \quad m = 1, 2, ..., \left(2^N - 1\right) \qquad (4.2.1.4)$$

i, j as previously defined. Once the initial set of activities is defined, only those which are ***valid permutations of active events*** are kept and then checked for compatibility as justified in the sequel. Moreover, after the first iteration, only the total probability of the string is of importance, since it represents the combined probability of the activity under consideration.

6. Define the set of augmented activities B_{jmr} (augmented strings Y_{jmr}) associated with each u_j obtained by inserting strings of ***valid active repetitive events*** within appropriate positions of the valid activities A_{jm} (strings X_{jm}) where r indicates the rth string of valid active repetitive events. The corresponding probability of an augmented activity-string is defined as:

$$P(Y_{jmr}/u_j) = p(M_{jmr}/X_{jm})P(X_{jm}/u_j) \qquad (4.2.1.5)$$

$$\sum_m \sum_r p(M_{jmr}/X_{jm}) = 1 \ \forall_j \qquad (4.2.1.6)$$

where M_{jmr} is the rth permutation (mask) matrix, or augmented mask matrix, associated with X_{jm}, $p(M_{jmr}/X_{jm})$ denotes the probability of the rth string of valid active repetitive events associated with X_{jm} and defined via the permutation matrix M_{jmr} (or augmented permutation matrix), and r is defined accordingly based on the specific application problem.

Table 1. Notations

E_t =	{set of primitive events in the task domain}
E_{nr} =	{set of non repetitive primitive events}

E_r = {set of repetitive primitive events}

E_c = {set of active non repetitive events associated with a u_j}

E_{snr} = {set of non repetitive events allowed to start an activity associated with a u_j}

E_{sr} = {set of repetitive events allowed to start a plan}

E_{end} = {set of non repetitive primitive events allowed to end a plan}

E_{cr} = {list of sets of critical non repetitive events associated with a u_j}

E_{comp} = {list of compatible pairs of primitive events}

E_{unw} = {list of unwanted precedence pairs of primitive events}

E_{ord} = {list of valid repetitive orderings}

Set Theoretic Relations

$E_{snr} \subset E_c \subset E_{nr}$

$E_{end} \subset E_c \subset E_{nr}$

$E_{sr} \subset E_r$

$E_{nr} \cup E_r \cup E_c \cup E_{snr} \cup E_{sr} \cup E_{end} = E_t$

$E_{comp} \subset (E_t \times E_t)$

$E_{unw} \subset (E_t \times E_t)$

7. Define the set of complete plans $Z_{jm\nu}$ associated with each u_j which is a *subset* of B_{jmr} and whose members are augmented activities satisfying some criteria for completion (described below).

The maximum number of the generated activities-strings A_{jm} (X_{jm}) is obtained when one considers that all events are active and all generated activities correspond to valid permutations. In reality, those numbers are reduced considerably due to compatibility tests justified in the next section. For example, when one considers only $(N - L)$ active non repetitive events, the initial number of activities A_{jm} (strings X_{jm}) becomes $\left(2^{(N-L)} - 1\right)$. Although permutations do increase this number, it is reduced considerably when one considers only compatible valid permutations. Moreover, when it comes to calculating the probability of an activity/string (see (4.2.1.4)), it is essential to understand that there is an alternative and actually easier way to evaluate these probabilities. Based on the principle of indifference, assign the initial a priori probability of the activity/string to be $1/\left(2^N - 1\right)$, and then update this probability based on learning. Augmented activities-strings B_{jmr} (Y_{jmr}) are generated only from valid activities-strings A_{jm} (X_{jm}) (compatible permutations) and then checked for compatibility, too. An alternative to the above is to consider immediately valid permutations of active non repetitive events [236]. The proposed approach does not restrict the generality or flexibility of the algorithm; it simply utilizes to steps to reduce the memory storage requirements. Storage requirements become enormous when no constraints are imposed on the string generation [237]. Although in reality this is the case, the organization level algorithm is described in its most general form, with minimum constraints attached.

The probability distributions depend critically on the a priori knowledge-information available. However, in order to avoid biasness and according to the principle of indifference, it is highly recommended to consider initially uniform distributions. Uniform distributions reflect a state of maximum uncertainty. Thus, the assumption is very valid because every system starts from the state of maximum uncertainty and reduces it as it continues to function. An alternative approach would be to derive the initial distributions based on Jayne's maximum entropy principle [93, 94].

Functions of the Organization Level

Functions are internal operations on the activities and ordered activities. As such, they are defined in their right order within the organizer.

Machine Reasoning (MR), (passive), is the association of the (classified) compiled input command, $u_j \epsilon U$, with a set of pertinent activities A_{jm} and production rules and procedures constituting the inference engine of the system (active reasoning). The latter procedures and rules have been assumed a priori for the work in this monograph (passive reasoning).

Machine Planning (MP), is the formulation of complete and compatible ordered activities to execute the requested job. Incomplete and incompatible ordered activities are rejected during this function to reduce the dimensionality of the model. Machine planning contains ordering of the active non repetitive primitive events within the activities, rejection of incompatible ordered activities, insertion of valid sequences of repetitive events (only) in the strings of compatible ordered non repetitive events, checks for completeness and organization of complete plans.

Machine Decision Making (MDM), is the selection of the complete and compatible ordered activity with the highest associated probability of success.

Machine Learning/Feedback (MFB), is calculation of the different individual and accrued cost functions associated with the execution of a requested job, and upgrading of the respective probabilities through learning algorithms. It is a key function of the Intelligent Robotic System because it provides the means to incorporate knowledge and task knowledge into the system (implicit knowledge). The machine feedback function is performed after the completion of the requested job and communication of selective feedback from the lower to the higher levels.

Machine Memory Exchange (MME), is retrieval, storage and upgrading of information from the long term memory of the organizer. Retrieval is performed during machine reasoning and machine planning, storage and upgrading after machine decision making and actual execution of the requested job.

Entropies and transmission of information evaluate the uncertainties related to these functions and indicate their interrelatedness.

As evidenced by the definitions, each one of the organization level functions is related to a certain objective. This objective may be thought of as a *local goal*, the satisfaction of which allows the system to proceed to the next function. Based on this, and without considering any details of how each function is performed, the local goal of machine reasoning is the association of the compiled command with a set of pertinent activities and procedures, that of machine planning is the derivation of complete plan scenarios, that of machine decision making is the selection of the best plan based on some specific criterion, that of machine learning/feedback is the reward or punishment of certain plan scenarios, while the local goal of machine long term memory exchange is to update the knowledge base of the organizer. The first three functions are associated with top-down local goals while the last two with bottom-up local goals.

Operational Procedures in the Organization Level

Any hierarchical system, and in particular an Intelligent Robotic System, is viewed as an entity of transforming user requested jobs into a sequence of specific actions.

Once a user command is issued (typed in, speech) it is preprocessed for classification and translation in machine code that is easily understood by the organization level of the system. Since these commands are general and sometimes fuzzy, it is expected that the system itself will develop suitable plan scenarios from its task domain to execute the user requested job. The formulation and organization of these plans requires the previously defined five functions to be performed within the organizer.

A plan is defined as a sequence of relevant and compatible non repetitive and repetitive active events-actions from the task domain E_t, appropriately arranged for sequential execution. The non repetitive events provide the framework for the plan(s) while the repetitive ones are necessary for completeness.

The relevance of primitive events is demonstrated via the concept of *criticality of events* introduced to ensure that *all necessary* non repetitive events from the task domain are present within all possible plan scenarios associated with a specific command. This concept is viewed as a *mapping* between the classified command u_j and a critical subset of active non repetitives defined through the initialization procedures of the system and utilized during machine reasoning. It should be noted, however, that not all active non repetitive primitive events associated with a compiled command are necessarily critical. This restriction is essential to reduce the complexity of the problem and keep the number of activities A_{jm} (strings X_{jm}) manageable.

The compatibility of primitive events is determined in terms of compatible pairs of them of the form $(e_k\ e_l)$, where e_k precedes e_l but not vice versa. Moreover, in order to avoid the derivation of compatible but irrelevant sequences of tasks (plan scenarios) associated with a specific command (quite possible as proved in [236]), we restrict the manner in which primitive actions are allowed to happen after others within a plan, at the expense of fewer activities and, thus, plans generated. We introduce the *constraint of precedence* to avoid unwanted sequences of events. The constraint, specified as a pair of events $(e_\mu\ e_\psi)$ rejects all activities-strings of non repetitive events in which e_ψ comes after e_μ in the string. These two tests are sufficient to generate compatible sequences of primitive events [236]. In addition to the above, to keep the dimensionality and complexity of the plan scenarios within acceptable limits, we propose the constraint of a *valid repetitive ordering* defined as a string of one or more *active repetitive* events satisfying the compatibility test described above. Valid repetitive orderings are those orderings whose first and last repetitive events are compatible with their non repetitive events to their left and right respectively, i.e.,$(e_\mu * * * \ldots * e_\psi)$ where the *'s indicate a valid repetitive ordering, the e's non repetitive ones and

$(e_\mu\,*)$ and $(*\,e_\psi)$ are compatible pairs.

The above constraints have mainly been *imposed on strings of non repetitive primitives only.* This seems to be natural and logical, since the non repetitive events establish the plan framework and the repetitive ones add the details. Thus, *the valid repetitive orderings of active repetitive primitive events are generated and inserted between adjacent non repetitive active ones.* The operational procedures of the organizer are described next.

Rule 1: The user commands $c_n, n = 1, 2, ..., M$, are considered to be independent of each other. The classified compiled input commands to the organization level u_j, $j = 1, 2, ..., M$, are also considered to be independent of each other. But due to classification, the selection of a (classified) compiled input command u_j depends on the user command c_n. Since the (classified) compiled input command u_j and the primitive events e_i are assumed to be probabilistically independent, different activities are independent of each other.

Rule 2: A string of events satisfies the compatibility test if for every (repetitive or non repetitive) event e_i in the string, e_i can follow the event immediately to its left in the string, and e_i can come before the event immediately to its right. For example, event e_i can come before event e_z if the form $(e_i\, e_z)$ is in the list of compatible pairs. Similarly, e_i can follow e_z if the form $(e_z\, e_i)$ is in the list of compatible pairs. An exception to this rule is the first and last event of the string, in which case additional information has to be supplied as to whether a particular repetitive event can start the string or a non repetitive event can end a string. The above condition follows from Rule 7 which states that a complete plan should start with a repetitive and end with a non repetitive event, as explained below.

Rule 3. A string of non repetitive events satisfies the precedence test if for every event e_i in the string, e_i can precede all other events that follow, though not necessarily immediately. Event e_i can precede event e_z if the pair $(e_i\, e_z)$ is *not* in the list of unwanted precedence pairs. An unwanted precedence pair $(e_z\ e_i)$ implies that e_z cannot precede e_i. This is another way to check compatibility.

Rule 4: The relative order of non repetitive events is crucial for the formulation of activities. If there is a maximum number of $(N - L)$ non repetitive events (critical or not) associated with an input command and q $(q \leq (N - L))$ is the minimum number of critical non repetitive events related to the input command and determined during the initialization of the system, then the activities associated with the particular command may be generated by considering all the valid permutations of these events taken l at a time, $l = 1, ..., (N - L)$. It follows that *an activity has at least* q *but not more than* $(N - L)$ *non repetitive events.* A valid permutation meets the following conditions:

1. Satisfies the compatibility test.
2. Satisfies the precedence test.
3. The first event is allowed to start the activity.
4. The last event is allowed to end the activity, and,
5. Includes the necessary critical events.

To specify q, assume there are several basic operations to execute a command, with critical (active non repetitive) events:

$$\{e_1, e_2, ..., e_{s1}\},\ \{e_3, e_5, ...e_{s2}\}, ...\{e_7, e_9, ...e_{sv}\} \qquad (4.2.3.7)$$

corresponding to each basic operation. Then $q = \min\{s1, s2, ..., sv\}$.

Rule 5: Let L be the number of repetitive events in the task domain. A valid ordering of repetitive events meets the following conditions:

1. It has at least one but not more than L repetitive events, each event in the string being unique.
2. The first event of the string can follow at least one non repetitive event in the task domain or it can start the plan, and,
3. The last event of that same string can come before at least one non repetitive event, and the ordering satisfies the compatibility test.

If the ordering consists of a single repetitive event, then the first and last events are, by definition, the same.

Rule 6. Formulation of augmented activities is performed by inserting valid strings of repetitive events in all different positions between non repetitives in each activity. In general, a complete plan represented as an ordered list of events will have the form $(*e_i * e_j * .... * e_z)$ where the e's are the non repetitive events and the *'s represent valid repetitive orderings whose first and last events are compatible with adjacent non repetitive events (with the exception of the first *). Considering that the original activity used to derive the complete plan is $(e_i e_j ... e_z)$, an augmented activity will have any of the following forms:

$$\begin{array}{c} (*e_i e_j ... e_z) \\ \bullet \\ \bullet \\ \bullet \\ (*e_i * e_j * ... * e_z) \end{array} \tag{4.2.3.8}$$

obtained as described above.

Rule 7. A complete plan is an augmented activity that:

1. It begins with a repetitive event which can start the plan and ends with a non repetitive event.
2. Contains at least one repetitive event between non repetitive events, and,
3. Satisfies all compatibility tests.

The simplest plan is one that starts with a repetitive event followed by a non repetitive event.

Rule 8. The generated plans are coupled with the (classified) compiled input commands u_j, because their corresponding probabilities are conditioned on the received command u_j. This coupling is used to provide further information about the formulation of meaningful strings to complete the requested job.

All pertinent information is stored in the long term memory of the organizer at specific addresses. This information is accessed when a specific user command is issued.

The Mathematical Model of the Input Classifier

A typed in user command transmitted via a communication channel is corrupted by noise. The output waveform from the communication channel is the input to the classifier. Every user command c_n, may be classified into any one of the M compiled commands u_j. Since there are M different typed in commands and M different compiled commands, the classifier is faced with M^2 alternative decisions each time a user command c_n is issued. M of these decisions (classifications) are correct (i.e., (u_j < − > $c_j, j = 1, 2, ..., M$), while $M^2 - M = M(M-1)$ are misclassifications. A cost is assigned to each alternative decision according to Bayes decision theoretic approach, and a risk function is calculated for each decision [56, 64, 245]. Minimization of the risk function yields the best possible decision (classification) u_j. This u_j is the input command to the organization level. The mathematical formulation of this problem is as follows:

Consider that each of the M typed in user commands c_n, $n = 1, 2, ..., M$, with a priori probability distribution function $p(c_n)$, is transmitted via the communication channel as an electric signal (i.e. a voltage) of constant value v_n. This signal is corrupted by additive noise within the channel. The output waveform (from the channel) is sampled at constant time intervals to obtain Q samples $\omega_1, \omega_2, ..., \omega_Q$, as shown in Figure 4.1. With this, each user command c_n maps into a point in the Q-dimensional observation space. Each noise sample is considered to be a zero-mean Gaussian random variable q with variance σ^2. The noise samples at various instants are independent random variables and are independent of the user commands issued, with corresponding probability distributions:

$$p_{qi}(x) = \left(1/(2\pi)^{1/2}\sigma\right) \exp\left(-x^2/2\sigma^2\right) \qquad (4.2.4.9)$$

because the noise samples are Gaussian. Thus, the observations under

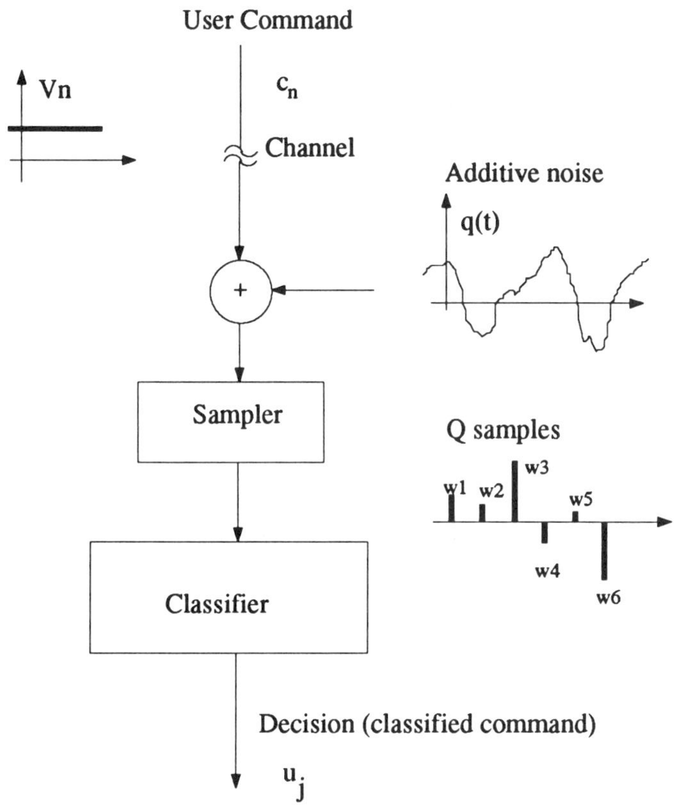

Figure 4.1 The Mathematical Model for the Classification Procedure

each user command are:

$$\omega_i = v_n + q_i; \; i = 1, 2, ..., Q, \text{ for all } n = 1, 2, ..., M \tag{4.2.4.10}$$

where ω_i is each observed sample.

The probability distribution of ω_i conditioned on each c_n is:

$$p(\omega_i/c_n) = p_{q_i}(\omega_i - v_n) = \left(1/(2\pi)^{1/2}\sigma\right) \exp\left[-(\omega_i - v_n)^2/2\sigma^2\right] \tag{4.2.4.11}$$

But the noise samples are statistically independent, thus the joint probability distribution of ω is simply the product of the individual probability distributions:

$$p(\omega/c_n) = \prod_{i=1}^{Q} \left(1/(2\pi)^{1/2}\sigma\right) \exp\left[-(\omega_i - v_n)^2/2\sigma^2\right]; \text{ all n's} \tag{4.2.4.12}$$

By Bayes rule, the a posteriori probability distribution function $p(c_n/\omega)$ that the user command is c_n based on the observation ω is:

$$p(c_n/\omega) = p(\omega/c_n)p(c_n)/p(\omega) \tag{4.2.4.13}$$

where:

$$p(\omega) = \sum_{n=1}^{M} p(\omega/c_n)p(c_n) \tag{4.2.4.14}$$

The conditional risk $R(u_j/\omega)$ associated with the decision that the user command is classified as u_j, given ω is [56]:

$$R(u_j/\omega) = \sum_{n=1}^{M} \lambda(u_j/c_n)p(c_n/\omega); \text{ for all j's} \tag{4.2.4.15}$$

where for simplicity:

$$\lambda(u_j/c_n) = \begin{cases} 0 \; ; \; j = n \\ 1 \; ; \; j \neq n \end{cases} \quad j, n = 1, 2, ..., M \tag{4.2.4.16}$$

a symmetrical loss function. Therefore, the conditional risk function may be rewritten as:

$$R(u_j/\omega) = \sum_{n \neq j} p(c_n/\omega) = 1 - p(c_j/\omega) \tag{4.2.4.17}$$

where $p(c_j/\omega)$ is the conditional probability that u_j is a correct classification.

Hence, the classification of a user command c_n into one of M (classified) compiled commands u_j is performed by defining a set of discriminant functions:

$$g_j(\omega) = -R(u_j/\omega) \tag{4.2.4.18}$$

and deciding that the compiled command will be u_j if:

$$g_j(\omega) > g_i(\omega) \text{ for all } j \neq i \tag{4.2.4.19}$$

For this particular case of minimum-error-rate classification, the symmetrical loss function is:

$$g_j(\omega) = p(c_j/\omega) \tag{4.2.4.20}$$

so that the maximum discriminant function corresponds to the maximum a posteriori probability. Therefore, for minimum-error-rate decide u_j if $p(c_j/\omega) > p(c_i/\omega)$ for all $j \neq i$. The block diagram of such a pattern classifier is shown in Figure 4.2.

The output of the classifier model is the input command u_j to the organization level. Its probability distribution function is defined to be:

$$p(u_j) = \max_j p(c_j/\omega) \; ; \; j = 1, 2, ..., M \tag{4.2.4.21}$$

the maximum a posteriori probability that minimizes the risk of a misclassification.

The algorithm of the organization level is described next. The most general description is presented. The operational procedures simplify the complexity of plans generation when imposed on the organizer functions. Given an application problem, specific constraints and assumptions apply as justified in chapter 6.

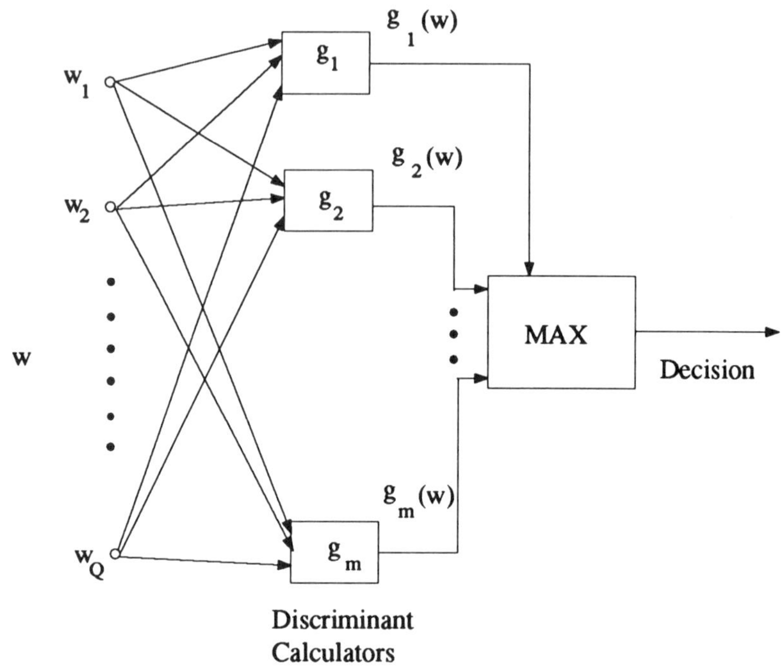

Figure 4.2 The Pattern Classifier Model

The Organization Level Algorithm

Step 1: System Input

An a priori probability distribution function, $p(c_n)$, is assumed for each user command:

$$p(c_n) = \nu_n/\nu_T \ ; \ n\epsilon\{1,2,...,M\} \tag{4.2.5.22}$$

$$\sum_{n=1}^{M} \nu_n = \nu_T \tag{4.2.5.23}$$

where ν_T is the total number of appearance of all user commands, and ν_n the relative number the command c_n has appeared.

The (classified) compiled input command to the organization level u_j has a probability distribution function given by (4.2.4.21). An explicit expression for $p(u_j)$ is obtained by substitution of (4.2.4.12), (4.2.4.14) and (4.2.5.22) to (4.2.4.13) and maximization according to (4.2.4.21). This gives:

$$p(u_j) = \max_j \frac{(\nu_j/\nu_T)\prod_{i=1}^{Q}\left(1/(2\pi)^{1/2}\sigma\right)\exp\left[-(\omega_i - v_j)^2/2\sigma^2\right]}{\sum_{n=1}^{M}(\nu_n/\nu_T)\prod_{i=1}^{Q}\left(1/(2\pi)^{1/2}\sigma\right)\exp\left[-(\omega_i - v_n)^2/2\sigma^2\right]}$$

$$j = 1,2,...,M \tag{4.2.5.24}$$

The denominator of (4.2.5.24) is the normalizing factor. Hence, maximization of the numerator is sufficient to yield the maximum $p(c_j/\omega)$. From (4.2.5.24) it is clear that each u_j depends on the user command c_n. It is also important to clarify that a noiseless transmission and an ideal classifier simply "translates" the user command c_j to u_j (so that it is understandable by the Intelligent Robotic System), with associated probability $p(u_j) = p(c_j)$.

The uncertainties of the user and compiled commands are given in terms of the entropy functions:

$$\begin{aligned} H(c_n) &= -p(c_n) \log p(c_n) \\ H(u_j) &= -p(u_j) \log p(u_j) \end{aligned} \tag{4.2.5.25}$$

The total entropies associated with all user and (classified) compiled commands are (Rule 1):

$$\begin{aligned} H(C) &= \sum_{n=1}^{M} H(c_n) \\ H(U/C) &= \sum_{n=1}^{M} H(u_j) \end{aligned} \tag{4.2.5.26}$$

and the transmission of information between two different user commands n and n' or compiled input commands j and j' is zero, i.e.:

$$\begin{aligned} T(c_n : c_{n'}) &= 0\,;\, n \neq n' \\ T(u_j : u_{j'}) &= 0\,;\, j \neq j' \end{aligned} \tag{4.2.5.27}$$

which also implies that:

$$\begin{aligned} H(c_n/c_{n'}) &= H(c_n)\,;\, n \neq n' \\ H(u_j/u_{j'}) &= H(u_j)\,;\, j \neq j' \end{aligned} \tag{4.2.5.28}$$

***Step 2:* Machine Reasoning (Passive)**

Once the compiled command u_j is known, an initial set of (maximum) $\left(2^N - 1\right)$ pertinent activities strings of active/inactive primitives is associated with it. Each activity A_{jm} is represented by a string of binary valued random variables $X_{jm} = \{x_1, x_2, ..., x_N\}_m$ corresponding to the mth activity and jth compiled command. Each string is assigned a probability (as previously described):

$$\begin{aligned} &Prob\,(A_{jm}/u_j) = P(X_{jm}/u_j) = P_{jm} \\ &j\,\epsilon\{1,2,...,M\},\, m = 1,2,...,\left(2^N - 1\right) \end{aligned} \tag{4.2.5.29}$$

(which is initially equal to $P_{jm} = 1/(2^N - 1)$) and it is updated by the learning algorithm described in ***Step 5***.

The entropy associated with a particular activity A_{jm} (string of binary valued random variables X_{jm}) conditioned on the (classified) compiled input u_j is:

$$H(X_{jm}/u_j) = -P_{jm} \log P_{jm} \tag{4.2.5.30}$$

The total entropy of a set of pertinent activities (strings) associated with one (classified) compiled command u_j is given in terms of its corresponding probability vector $S_j = \{X_{j1}, X_{j2}, ..., X_{jm}\}; m = 1, 2, ...(2^N - 1)$:

$$H(S_j/u_j) = \sum_{k=1}^{2^N-1} H(X_{jm}/u_j); \; j \epsilon \{1, 2, ..., M\} \tag{4.2.5.31}$$

whereas the entropy of all activities (strings) associated with all (classified) compiled commands is:

$$H_R = \sum_{j=1}^{M} H(S_j/u_j) \tag{4.2.5.32}$$

which is the quantity that represents the total uncertainty during machine reasoning at a particular stage of the system.

Rules 1 and 2 also imply that:

$$\begin{aligned} T(X_{jk} : X_{jl}/u_j) &= 0; \; k \neq l \\ H(X_{jk} / X_{jl}, u_j) &= H(X_{jk}/u_j) \end{aligned} \tag{4.2.5.33}$$

where X_{jk}, X_{jl} are any two different strings of binary valued random variables corresponding to the activities A_{jk}, A_{jl} respectively.

Step 3: **Machine Planning**

Once the set of activities A_{jm} has been associated with the compiled command, based on the operational procedures, the following steps occur which actually simplify the algorithm complexity and reduce memory storage requirements.

Define the sets E_c, E_{cr} and E_{snr} associated with the specific u_j, as defined in Table 1. . Then:

Step 1. Generate all the permutations of E_c, Per$\{E_c\}$ (non repetitive events only) taken l at a time, l = 1, 2, ...,(N − L) and such that:

1. The first event in each permutation ϵ E_{snr}.
2. The last event in each permutation ϵ E_{end}, and,
3. Each permutation satisfies the compatibility test (Rule 2).

Step 2. Reject permutations that do not satisfy the precedence test (Rule 4).

Step 3. Retain only those permutations from Per$\{E_c\}$ that contain all the necessary critical events associated with the particular u_j, that is, $E_{cr} \subset$ Per$\{E_c\}$.

Step 4. Initialize the set of augmented activities of the form $((*e_g * ... * e_i * e_j ... e_z))$ where the e's are active non repetitive events.

Step 5. Determine all the valid repetitive orderings from E_{ord} that may be substituted in place of the leftmost *'s in the set of augmented activities using the following procedure:

1. For the very first * (before e_g), a valid repetitive ordering $(e_v, ..., e_\psi)$ may be substituted for, if $e_v \,\epsilon\, E_{sr}$ and $(e_\psi\ e_g) \,\epsilon\, E_{comp}$.
2. For the remaining *'s, a repetitive ordering of the above form may be substituted for the * between e_i and e_j if $(e_i\ e_v)$ and $(e_\psi\ e_j)$ ϵ E_{comp}.

Step 6. Let K be the number of the valid repetitive orderings that may be substituted for the first * encountered, starting from the left $(k = 1, ..., K)$. Then:

i) If $K \neq 0$, modify the current set of augmented activities to equal K copies of itself and then substitute each of the K repetitive orderings for the leftmost *'s in the current set of augmented activities. Hence, considering the valid orderings to be $y_1, y_2 ....$, the set of augmented activities will be transformed and expanded as follows:

$$\begin{array}{c} \{(y_i e_g * \dots * e_i * e_j \dots * e_l)\}, \\ \bullet \\ \bullet \\ \bullet \\ \{(y_k e_g * \dots * e_i * e_j * \dots e_l)\} \end{array} \tag{4.2.5.34}$$

Do Step 3 until no more *'s are remaining.

ii) *Otherwise,* discard all the augmented activities derived from the current activity and then go to *Step 5.*

Step 7. Store the current set of augmented activities in the long term memory.

Step 8. Repeat the process for the other activities.

The ordered and augmented ordered activities B_{jmr} (strings Y_{jmr}) may be obtained via permutation (mask) matrices. The basic procedure to obtain ordered activities (from simple activities) in general, is as follows:

The ordering of the primitive events $e_i\,(x_i)$ is obtained by rearranging the order of the active x_i's of the vectors X_{jm}'s. This is accomplished through transition matrices (masks) which permutate the x_i's. *A mask matrix is defined as the* $(N \times N)$ *matrix with elements 1's and 0's such that there is only one 1 in every row and column.* For each activity (string with k active primitive events) there are (maximum) $k!$ permutation masks and therefore $k!$ ordered activities (strings) may be obtained.

For illustration purposes, consider an activity (string) with active primitives $e_1e_2e_3$ $(k = N = 3)$ represented by the string $X_{jm} = x_1x_2x_3$. There are $3! = 6$ ordered activities (strings), assuming that all orderings are compatible, obtained from $e_1e_2e_3$ $(x_1x_2x_3)$ via the six

(3x3) permutation matrices:

$$\begin{bmatrix}1&0&0\\0&1&0\\0&0&1\end{bmatrix}, \begin{bmatrix}1&0&0\\0&0&1\\0&1&0\end{bmatrix}, \begin{bmatrix}0&1&0\\1&0&0\\0&0&1\end{bmatrix}$$
$$\begin{bmatrix}0&1&0\\0&0&1\\1&0&0\end{bmatrix}, \begin{bmatrix}0&0&1\\1&0&0\\0&1&0\end{bmatrix}, \begin{bmatrix}0&0&1\\0&1&0\\1&0&0\end{bmatrix} \quad (4.2.5.35)$$

The (compatible) ordered strings of binary valued random variables Y_{jmr} are obtained through:

$$Y_{jmr} = M_{jmr} X_{jm} \; ; \; m = 1,2,...,\left(2^N - 1\right) \qquad r = 1,2,...k! \quad (4.2.5.36)$$

where M_{jmr} is the *r*th permutation mask associated with the activity X_{jm}. The ordered strings of binary valued random variables for the previous example are:

$$\begin{aligned} Y_{jm1} &= y_1 y_2 y_3 = x_1 x_2 x_3 \\ Y_{jm2} &= y_1 y_2 y_3 = x_1 x_3 x_2 \\ Y_{jm3} &= y_1 y_2 y_3 = x_2 x_1 x_3 \end{aligned} \quad (4.2.5.37)$$

$$\begin{aligned} Y_{jm4} &= y_1 y_2 y_3 = x_2 x_3 x_1 \\ Y_{jm5} &= y_1 y_2 y_3 = x_3 x_1 x_2 \\ Y_{jm6} &= y_1 y_2 y_3 = x_3 x_2 x_1 \end{aligned} \quad (4.2.5.38)$$

The above technique may be expanded to allow for insertion of valid sequences of repetitive events. When this is the case, the permutation/mask matrices become non square matrices with several 1s in each row as analytically demonstrated in [237].

Of course, when compatibility is considered and all incompatible ordered strings are rejected, the total number is reduced very considerably.

Define $p(M_{jmr}/u_j)$ to be the probability of the *r*th compatible permutation mask used to obtain Y_{jmr} from X_{jm} given u_j. Then, the probability of the rth ordered string Y_{jmr} is:

$$P(Y_{jmr}/u_j) = p(M_{jmr}/u_j)P(X_{jm}/u_j)$$
$$\sum_m \sum_r p(M_{jmr}/u_j) = 1 \qquad (4.2.5.39)$$

Step 4: **Machine Decision Making**

Decision making is simply the choice of the most probable plan, defined as the plan corresponding to a given command with the highest probability of success.

After the selection of the most probable plan associated with a specific command u_j, this plan is communicated to the coordination level where a specific execution scenario is developed by that level. Plan execution follows. After the plan execution (successful or unsuccessful), feedback information is communicated from the coordination to the organization level. This is the function of machine learning, feedback, described in the sequel.

Step 5: **Machine Feedback**

Define the overall accrued cost J_c associated with the execution of specific plans where the particular form of J_c depends on the specific application problem.

Upgrade all pertinent probabilities using the general stochastic approximation learning algorithm:

$$p(t + 1/c_i) = p(t/c_i) + \beta_{i+1}[\xi - p(t/c_i)] \qquad (4.2.5.40)$$

where t is the iteration number, β_{i+1} is a sequence satisfying Dvoketcky's condition for convergence [185], $p(t + 1/c_i)$ is the corresponding probability of the command c_i, J' is the actual cost of execution of the formulated plan, and:

$$\xi = \begin{cases} 1; J' = \min J_c \\ 0; \text{otherwise} \end{cases} \qquad (4.2.5.41)$$

A more general case is to consider that the performance index is corrupted by the uncertainties of the system [76, 185]. If this is the case, performance estimates are first updated with the expression:

$$J(t + 1/c_i) = J(t/c_i) + \gamma_{i+1}[J_{obs}(t + 1/c_i) - J(t/c_i)] \tag{4.2.5.42}$$

where J_{obs} is the observed performance value, J is the performance estimate, γ_{i+1} is a sequence satisfying Dvoketcky's condition for convergence, and the J's obtained are used in the ξ equation above.

***Step 6:* Machine Memory Exchange**

This function involves retrieval, storage, and upgrading of information in the long term memory of the organizer. Once a compiled command has been recognized, a memory package is associated with it that contains relevant information. After the formulation of the plan and execution of the requested job, the upgraded information is stored in the organizer memory.

The best possible plan to execute the requested job is stored in a buffer accessible by the dispatcher of the coordination level. Due to the hierarchical control structure, planning is hierarchical, too. Each task in the organization level is decomposed into subtasks at the coordination level all the details associated with plan(s) execution are defined at the coordination level. Thus, it is the objective of the coordination level to formulate the actual control problem and decide how the plan will be executed.

All entropy and transmission of information equations are easily transformed to their corresponding rates based on the rate rule. Considering that the entropy is measured in information nats, the corresponding rate is given in information nats per step (iteration cycle).

An alternative model for the organization level is described next. This model is basically an expert system cell and it is suitable (and actually works better) for situations where the system has a good understanding of the different problems it is faced with. However, the expert system cell model does include a "probabilistic mode" for cases of very incomplete knowledge related to the user requested job, which is essentially the already described model in this section.

4.3 The Expert System Cell for the Organization Level

Modeling of the organization level based on expert system techniques is performed according to the separate identification of the various components of the robotic system organizer (RSO), namely, the organizer control system unit (OCSU) (simply used as a communications protocol to establish relative priorities among the different system parts), the input classifier (IC), the knowledge base (KB) and the inference engine (IE) [81]. This type of modeling approach is followed because modifications can be made within one or more of the organizer's components without any violations of the structure of the entire system [235]. The robotic system organizer is capable of performing functions related to:

1. User command classification.
2. Task selection and planning given a user requested job.
3. Formulation of complete task plans with or without interaction with a human operator.
4. Supervised and/or unsupervised learning functions.
5. Automatic update of information within the knowledge base (dynamic knowledge base), and,
6. Conduct simulation runs for output verification before the actual real time execution of a formulated plan.

The knowledge base is the part which contains all information related to the domain of problems an Intelligent Robotic System is designed for, while the inference engine operates upon it. Both the knowledge base and the inference engine are designed in such a way, that, although they are considered separate entities, they operate closely together as a whole. The size of the knowledge base, in which all necessary information about a specific problem domain is stored, is usually very large and is structured with some kind of syntactic format for efficient manipulation of the stored data [60, 81]. This syntactic format is often represented in terms of graph or table forms utilizing Group Theory. The inference engine is the active part which uses Artificial Intelligence techniques to perform specific functions on the knowledge base.

In general, given a user command, the organizer first preprocesses the command by classifying it into one of command types in a similar way as previously described. Then, a sequence of actions is created for that particular command either via forward chaining or probabilistically. The forward chaining method is used to generate the probable actions sequences when there is already some information stored in the knowledge base about the user command; otherwise, a probabilistic approach is followed automatically as previously described. Before creating a possible plan, many compatibility tests are also carried out to ensure that the generated plan is correct and complete. Finally, once a complete and compatible possible sequence is formulated, it is sent to the coordination level and from there to the execution level. According to the failure or success of the generated plan, determined via feedback, either more attempts are made to come up with alternative planning scenarios, or an update of the knowledge base is performed via learning algorithms [65, 66]. When all alternatives have been checked without a successful completion of the user requested job, the organizer automatically resorts to the user for help. By using supervised learning, additional basic task functions or command sequences are introduced, so that the organizer may carry out the desired user command properly.

Subsystem Structures and Functions

Input Classifier

The purpose of the input classifier is to interpret and classify a user command into one of several command types as previously described. If the received command is a "new" command in the sense that its "type" is different from the existing command types, a new category is generated to store the new command type. Misclassification situations cannot be completely avoided but they are reduced with the aid of a minimum error rate classification algorithm, as previously described.

The Knowledge Base

The knowledge base contains all relevant information associated with the user commands and the task domain functions. Since optimal modeling of the knowledge base and optimal access of the information stored in it are crucial to the system performance and efficiency, all information stored in the knowledge base is accessed in terms of specific addresses each of which resides in the system's memory and contains fixed information within an operation cycle. This is in accordance with the probabilistic model, too. Therefore, it is sufficient to consider that the knowledge base is composed of five different table-like structures, namely, *command table, command type table, instruction table, instruction type table,* and, *instruction set table.* Each table is stored separately in the knowledge base at prespecified address locations.

The command table contains the list of all predefined user command names, e.g., TURN-ON-CEILING-LIGHT, TURN-ON-DESK-LIGHT. The command type table contains all the different command types. For example, all TURN-ON commands are identified as one type. The instruction table contains the set of elementary organization level tasks/actions (equivalent to the repetitive and non repetitive events). The instruction type table is similar to the command type table; it stores the different types of available event functions. These functions are grouped and accessed when the appropriate coordinator is activated. Finally, the instruction set table contains existing overall plans (if any) corresponding to the user commands. These complete scenarios are already successfully executed scenarios.

In mathematical terms, the command table is defined as the set of user commands:

$$C = [c_1, c_2, , ..., c_M; \; M \text{ finite}] \qquad (4.3.1.43)$$

The command type table is then defined in terms of the sets $CT_1, CT_2, ..., CT_k$, k finite, where the CT_i's, $i \epsilon [1, k]$, are the different command types, TURN-ON, TURN-OFF, etc. Each CT_i is defined as:

$$CT_i = [c_k, c_l, ..., c_m] \text{ for } i\epsilon[1, k],\ k, l, m \,\epsilon\, [1, k], \quad (4.3.1.44)$$

$$\begin{aligned}(CT_i \cap CT_j) &= 0, \text{ for } i \neq j \text{ and } i, j \,\epsilon\, [1, k]\ CT_1 \cup \\ &CT_1 \cup ... \cup CT_k = C\end{aligned} \quad (4.3.1.45)$$

which guarantee that each user command can be of at most one command type. The instruction table is defined in terms of the e_j's which indicate the elementary organizer actions (repetitive and non repetitive primitive events), and the instruction type table in terms of the sets $ET_1, ET_2, ..., ET_p, p$ finite, with each $ET_p, p \,\epsilon\, [1, P]$, defined as:

$$ET_p = [e_q, e_s, ..., e_t], \text{ for } p\epsilon\, [1, P], \text{ and } q, s, t \,\epsilon\, [1, j], \text{ for } q \neq s \neq t \quad (4.3.1.46)$$

with:

$$ET_1 \cup ET_2 \cup ... \cup ET_p = E \quad (4.3.1.47)$$

This indicates that the same elementary event/action may be used by more than one coordinators during the formulation of the planning sequencies. The instruction set table is defined as:

$$IST = [P_1, P_2, ...P_y; y \text{ finite}] \quad (4.3.1.48)$$

where P is a task sequence composed of e_j's. Details related to the robotic system organizer are now presented:

Command Table: This table contains the set C of all predefined user commands. They are ordered according to their command name and type to increase the effectiveness of the search algorithm developed for classification purposes. The command table, set C, interacts with the command type table, set CT. The order of the command types within each CT_i follows the order in which each user command is stored in C. At the moment, the maximum number of user commands allowed is 100, but this can be expanded easily. Each record of the command table contains two fields, a **command name** field and a **ways** field. The command name field is a string of (maximum) 40 characters long which contains the actual command names. The ways field contains an integer from 0 to 9 indicating the number of **extra** different ways a particular command can be executed. This information is obtained from the instruction set table which, as mentioned above, contains all relevant plans associated with each user command. For example, an entry in the command table might read: TURN-ON-CEILING-LIGHT (1); this means that for the defined user command TURN-ON-CEILING-LIGHT, there are two plans (1 basic and 1 "extra") capable of executing that particular command as shown in Figure 4.3(a).

Command Type Table: This table contains all the different existing command type sets $CT_1, CT_2, ..., CT_k$. Each record of the table contains three fields: **command type, front** and **back** fields. The command type field contains the name of the command type and the front and back fields point to the beginning and the end of that particular command type within the command table. For example, an entry in this table might read: TURN-ON (1) (5); this means that all of the defined user commands starting from record 1 through record 5 of the command table are classified as command type TURN-ON, and these five user commands are arranged alphabetically. The command type field is also 40 characters long and can store a maximum of 20 different types of commands. These numbers, 40 characters and 20 different types, are arbitrarily chosen and may be changed according to the user specifications, provided they are within the limits of the system. A block diagram is shown in Figure 4.3(b).

Instruction Table: This table contains the list, set E, of all pre-defined basic instructions (actions) $\{e_1, e_2, ..., e_N\}$. These elementary instructions are considered to be the smallest piece of information that may be processed by the organizer. They are classified either as repetitive or non repetitive events as previously explained. A repetitive basic instruction is contained in all planning sequences regardless of command type to ensure completeness of the plan scenario. Non repetitive basic instructions belong to certain command types only. The reason for this restriction is because the non repetitive events/actions are the essential factor in classifying each user command. When a planning sequence is formulated and passed to the coordination level, these basic instructions are further decomposed into microinstructions which contain detailed information about the specific way each basic function is executed. Task functions are 50 characters in length, and a maximum number of 100 different instructions is allowed in this table. A block diagram is shown in Figure 4.3(c).

Instruction Type Table: Each instruction is classified according to the coordinators of the system; for an Intelligent Robotic System, the Vision System Coordinator (VSC), Motion System Coordinator (MSC), Gripper System Coordinator (GSC), and Sensory System Coordinator (SSC). There is a maximum number of ten allowed coordinators at the coordination level. This number may change according to the specific application problem. Each record set, ET_j, in this table contains the coordinator's name in the "instruction type" field. It has the similar **front** and **back** fields to the ones of the command type table which point to the beginning and the end of the instruction table, the instruction type is associated with. For example, an entry in the instruction type table might read: VSC (1) (3); this means that there are 3 basic functions associated with the VSC. Within each instruction type, the instructions are ordered alphabetically according to the instruction name for maximum efficiency when searching through the table. A block diagram is shown in Figure 4.3(d).

Instruction Set Table The instruction set table, *IST*, contains all successful sequences for the execution of pertinent user commands from the command table. At most 10 different possible instruction sets (plans) can be stored at once for a particular command; they start at location (*command number *10* to (*command number * 10+9)* within the *IST*. For example, if the user command TURN-ON-CEILING-LIGHT is located at the first record in the command table, then the corresponding task sequences will begin at location (1 * 10), or record 10, through a maximum of (1 * 10 + 9), or record 19, in the *IST*. The number of different successful planning scenarios associated with a particular command is indicated in the ways field of the command table as discussed earlier. Since most plans contain from twenty to thirty instructions within a task sequence, a maximum of 40 basic instructions for each task plan is allowed. Note that in this table, the first nine records located at 1 through 9 are not used at this time due to the specified structure that the knowledge base takes, but can be utilized in other ways to keep track of other pertinent system information as shown in Figure 4.3(e).

The Inference Engine

The inference engine of the organizer uses Artificial Intelligence techniques to manipulate the knowledge base efficiently. In general, the IE goes through four sequential stages, namely, *initialization, formulation, execution* and *finalization*. During initialization, the inference engine sets up its knowledge base structures and reads pertinent knowledge (information), if there is any, from secondary storage (large but slow) memory. Then it prompts the user by displaying the main function menu from which the user can perform one of the following:

1. Perform task planning functions with user specified commands (this enables the user to run task generating functions).
2. Add elementary functions or task plans to the knowledge base (this enables the user to build up, modify, or delete existing basis task functions or user commands), and,
3. Exit program (update the knowledge base files and system clean up).

I. Command Table

(1,2,...,40 in length)

	cmd name	ways
1	TURN_ON_CEILING-LIGHT	1
2	TURN_ON_DESK_LIGHT	2
	CLOSE_WINDOW	0
100		

(a)

II. Command Type Table

(1,2,...,40 in length)

	cmd type	front	back
1	TURN_ON	1	5
	TURN_OFF	6	10
	CLOSE	20	22
20			

(b)

III. Instruction Table

(1,2,...,40 in length)

	ins name	rep/nonrep (1/0)
1	SCAN	
2	IDENTIFY	
MOH INS	TURN	
100		

(c)

IV. Instruction Type Table

(1,2,...,40 in length)

	ins type	front	back
	VSC	1	3
	SSC	4	5
TYPES	GSC	13	20
10			

(d)

Figure 4.3 Knowledge Base Structures of the Organizer

V. Instruction Set Table

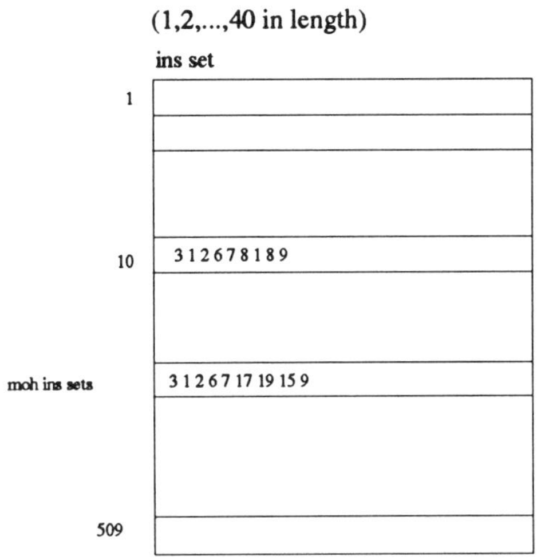

(e)

Figure 4.3 (Continued)

Selection 1 brings the program to the formulation stage. Once a command is issued by the user, it is first classified into one of the existing command types, or to a new command type (not previously existing). If a command exists, then the plan will automatically be retrieved form the knowledge base; else, the user will be asked to make a choice of whether a forward chaining or probabilistic approach should be used to generate the new plan. There are different advantages related to these methods, and depending on the particular command at hand and the working size of the knowledge base, one method might be better than the other. In general, if there is sufficient information stored in the knowledge base about the issued command, then the forward chaining method is used to come up with the correct task plan since it is faster, in terms of execution, than the probabilistic approach. On the other hand, if none or very little information exists in the knowledge base, then a probabilistic approach is followed to come up with complete task sequences as previously explained. Moreover, if the user command is of a new command type, then the probabilistic approach is chosen automatically, since the forward chaining method cannot be utilized due to the lack of knowledge related to that particular user command type.

Once the task plan is either retrieved or generated, the plan is then executed. This is done by passing the task sequence to the coordination level. Next, an evaluation of the task plan takes place which is in the form of a direct feedback from the different coordinators. If the desire task is performed successfully, then the program returns to its main menu and at the same time updates the knowledge base, otherwise, the organizer tries to reformulate the next plan either by backtracking or selecting the next probable plan until a successful task plan is generated or until all existing or probable plans have failed. If all plans do fail to perform that specific user command, then the last alternative taken by the organizer is to ask the user for help via a teach box.

The teach box provides the user with the ability to add a basic task function, a new user command, or to revise an existing command task sequence. Once the necessary knowledge is added to the database, the organizer tries to regenerate new task sequences for that specific user command.

A detailed flow chart of the expert system cell is shown in Figure 4.4.

4.4 The Coordination Level

The coordination level is an intermediate structure serving as an interface between the organization and the execution level. It is essential for dispatching organizational information to the execution level. *Its objective is the actual formulation of the control problem associated with the most probable complete and compatible plan formulated by the organization level that will execute in real time the requested job.* This includes selection of one among alternative plan scripts that accomplish the same job in different ways according to the constraints imposed by the workspace model and timing requirements.

The coordination level is composed of a specified number of coordinators of fixed structure. Specific hardware (execution devices) from the execution level is associated with each coordinator. These execution devices execute well defined tasks when a command is issued to them by their corresponding coordinator.

The major advantage which results from this association is that the individual functions of each coordinator may be defined a priori (during the design phase of the system) because they are considered to be unmodifiable with time. Thus, they are assumed to be deterministic functions since the number of parameters involved in each one of them is also prespecified.

This structure implies that the coordination level does not have any reasoning capabilities like the organizer. Its intelligence is related to its ability on how to execute the organizer plan in the best possible way based on past experience and real time constraints associated with the workspace environment.

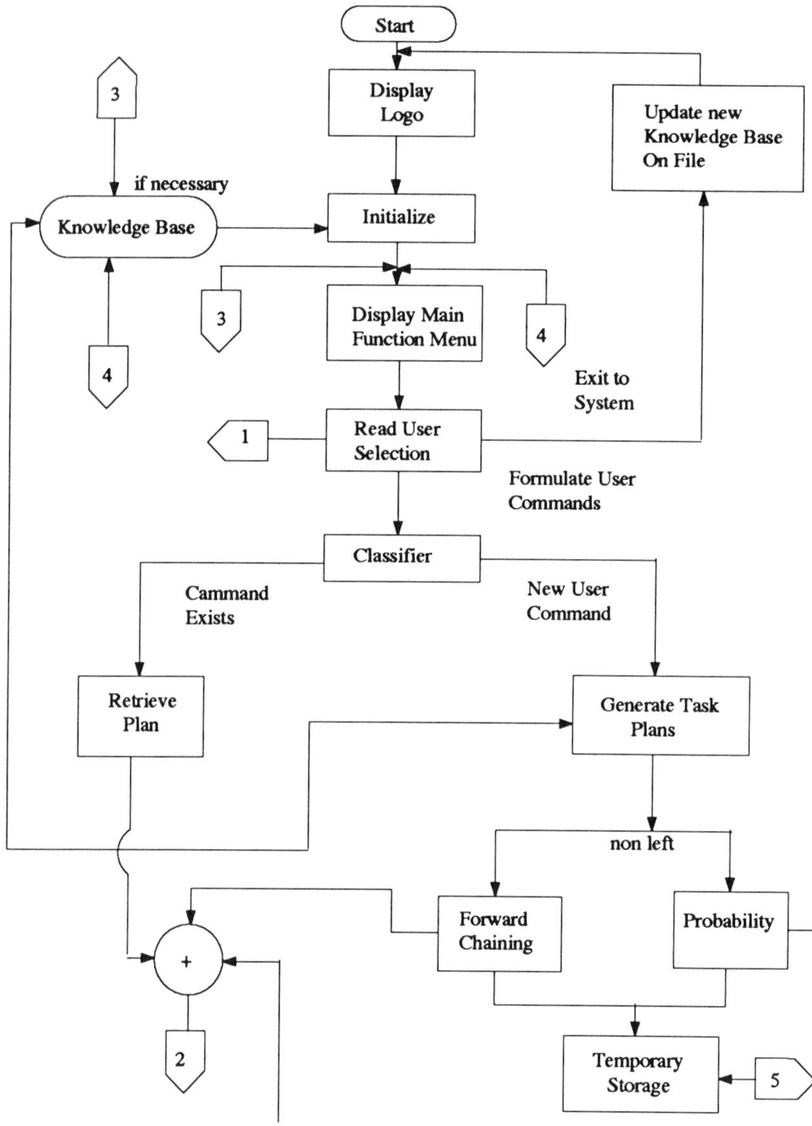

Figure 4.4 Main Program Flowchart

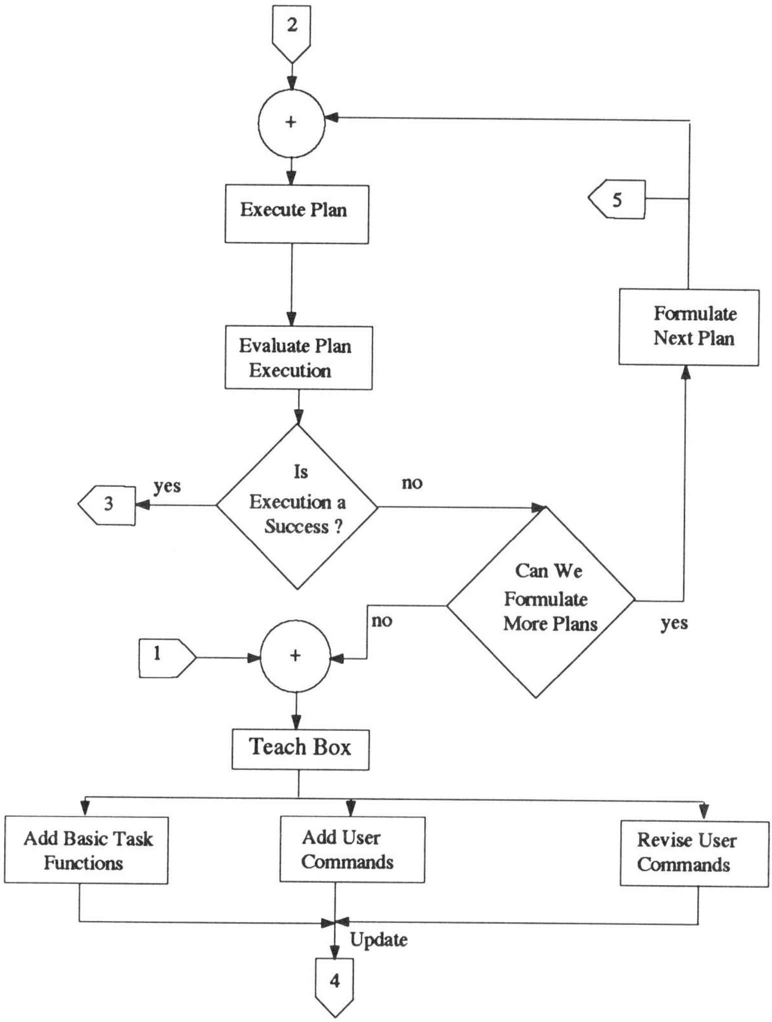

Figure 4.4 (Continued)

The coordination level involves decision making associated with specific knowledge (information) processing based on the already formulated plan. The flexibility of the coordination level and its ability to formulate alternate real time scenarios is influenced by the variable structure of the dispatcher which is dictated by the organization level.

All tasks generated at the organization level are decomposed into subtasks within the coordination level. All information related to task and subtask completion/execution is stored within the dispatcher memory and passed to the different coordinators when necessary. Although the different coordinators do not communicate directly with each other, due to the hierarchical nature of planning, the task decomposition is both *temporal* and *spatial.* Temporal decomposition corresponds to sequential tasks/subtasks while spatial decomposition corresponds to tasks/subtasks executed by more than one coordinators.

The functions of the coordination level are defined in terms of the individual functions of the different coordinators of an Intelligent Robotic System. These coordinators are shown in Figure 4.5 and are defined to be:

1. The Vision System Coordinator (VSC).
2. The Sensor System Coordinator (SSC).
3. The Motion System Coordinator(s) (MSC), and,
4. The Gripper(s) System Coordinator(s) (GSC).

It is important to clarify at this point that we consider the Vision System Coordinator as a separate coordinator and not as a part of the Sensor System Coordinator. The main reasons for this distinction are: First, machine and robot vision has become a very important component in modern robotic systems and robot vision systems are studied and treated separately from all other types of external (and internal) sensors. Second, the hardware associated with the Vision System Coordinator is different from the one associated with the Sensor System Coordinator in the sense that each non visual type of sensor has different operational requirements and specifications.

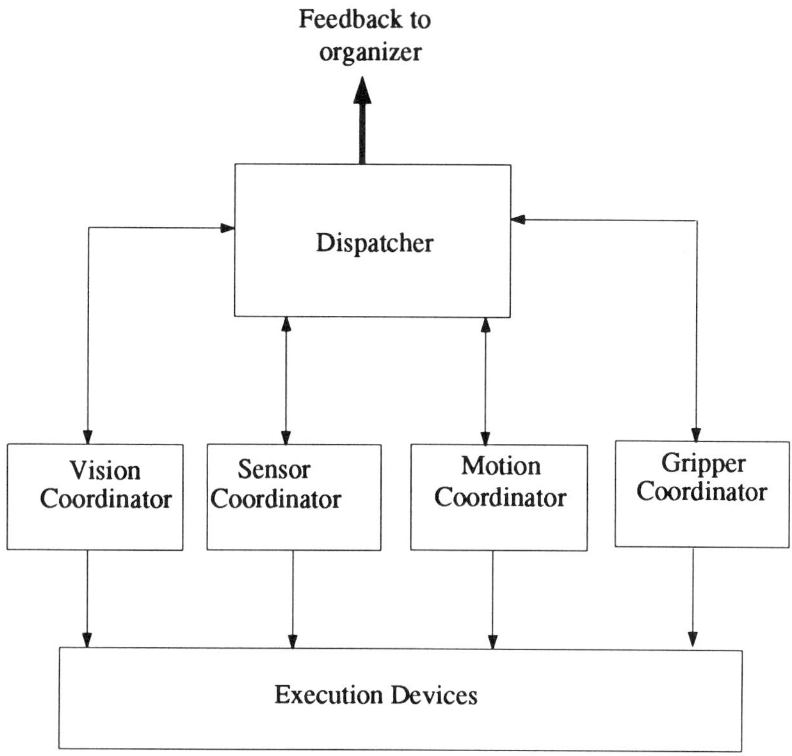

Figure 4.5 Block Diagram of the Coordination Level of an Intelligent Robotic System

Each coordinator when accessed, performs a prespecified number of different functions. A cost is assigned to each individual function. An accrued cost is associated with the operation of each coordinator. An overall accrued cost is calculated in terms of the weighted sum of the accrued costs of the coordinators after the execution of the requested job. This cost is communicated to the organizer *after the completion of the requested job* and is used to upgrade the information stored in the long term memory of the organization level. This feedback information (which is sent from the coordination to the organization level after the completion of the requested job) is *off line* feedback information. On the other hand, feedback information is communicated to the coordination level from the execution level *during the execution of the requested job.* Each coordinator when accessed issues a number of commands to its associated execution devices (at the execution level). During the execution and upon completion of the issued commands feedback information is received by the coordinator and is stored in the dispatcher. This information is used by other coordinators if necessary, and also to calculate the individual, accrued and overall accrued costs related to the coordination level. Therefore, the feedback information from the execution to the coordination level is *real time, on line* feedback information. More details about the feedback mechanism are given in the corresponding sections where the functions of each coordinator are explained.

The functions of the different coordinators are explained in the sequel. A description of their associated execution devices from the execution level is also presented.

The Functions of the Vision System Coordinator

The purpose of the Vision System Coordinator, the structure of which is shown in Figure 4.6, is to identify objects and/or tools, collect and provide information about their location, position and orientation in the workspace environment. When this coordinator is activated, it issues a number of different commands to its associated execution devices (at the execution level). These devices are the components of any vision system. The figure shows only cameras but there may also be range finders and laser spreaders. The commands issued to these execution devices (at the execution level) are:

1. Command a scanning of the environment workspace and collect data about objects and tools in the workspace.
2. Command an interpretation of the collected data to determine if an object or tool (necessary to execute a specific task) is present in the workspace, and,
3. Command a determination of the location, position and orientation of an object and/or tool.

The camera (or cameras) takes pictures of the objects and tools in the workspace and transfers their images to the memory of the Vision System Coordinator through the frame unit and monitor. Recognition of a particular object or tool is accomplished via Pattern Recognition and Scene Analysis Techniques [56]. The general recognition procedure is as follows: The object and tool images are stored a priori in the memory of the Vision System Coordinator. This information is considered sufficient to generate and recognize images from pictures taken from any optical angle in the workspace. In order to recognize a particular object or tool, its picture image is compared with the object and tool images already stored in the memory to choose the best matched one. Once an object or tool has been recognized, further information about its location (position and/or orientation) is provided (if and when necessary) to the Vision System Coordinator from the camera(s).

The accrued cost function associated with the operation of this coordinator is determined in terms of two individual cost functions related to the operations of the associated execution devices. One cost function is associated with the object or tool recognition procedure and the other with the determination of the location of an object or tool.

The cost function associated with the object or tool recognition procedure is related to the uncertainty involved in such a procedure.

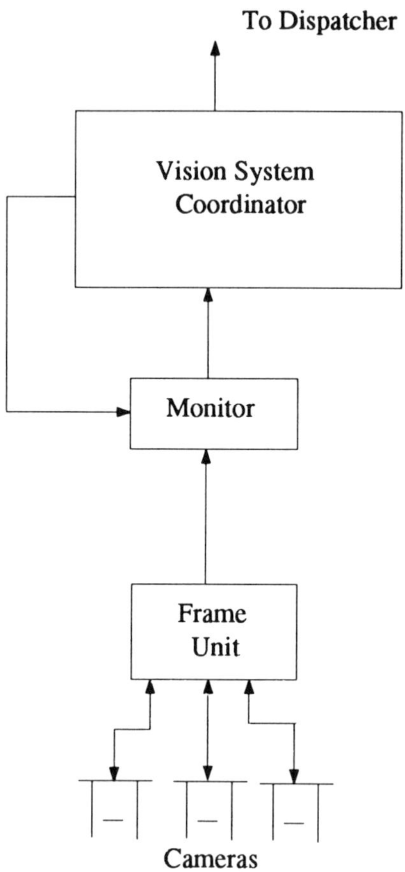

Figure 4.6 The Vision System Coordinator and Execution Devices

This uncertainty is a function of the geometry and symmetry of the object or tool, as well as of the optical angle from which a picture is taken. Furthermore, this optical angle is a function of the position and orientation of the object or tool. Therefore, given an object or tool with known geometry and symmetry, the uncertainty associated with its recognition procedure may be expressed in terms of its positional and orientational entropy:

$$J_1 = J(H(Q)) = H_{QP} + H_{QO} \tag{4.4.1.49}$$

where H_{QP}, H_{QO}, respectively represent the positional and orientational entropies of the object or tool [179].

The probability distribution functions for the position and orientation of an object or tool are derived in terms of its location in the 3-D Cartesian space: The position of an object or tool may be described by its cartesian coordinates (x, y, z) relative to an origin O. The orientation of an object or tool centered at O is most conveniently described by Euler angle rotations α, β, γ. Thus, uncertainty in position and orientation may be described by the joint probability distribution function $P(x, y, z, \alpha, \beta, \gamma)$ over the joint ensemble. A large number of relationships may be developed around this distribution. A widely used distribution in application problems considers that orientations are interdependent but independent of positions and vice versa, i.e.:

$$P(x, y, z, \alpha, \beta, \gamma) = P(x, y, z)\, P(\alpha, \beta, \gamma) \tag{4.4.1.50}$$

The orientation probability distribution is the most important because it is used as a measure of the complexity of an object or tool with respect to symmetry properties. Highly symmetric objects or tools are generally less complex and easier to recognize, i.e., a sphere [179]. The particular form of the probability distribution functions depends on the shape of the specific objects or tools related to an application problem. A more detailed analysis may be found in [179].

The cost function associated with the determination of the position of an object or tool is related to the uncertainty about the position of the object or tool. The true position in the workspace is unknown. What is known, are measurements of the position provided by the camera(s). These measurements introduce an error. Assume for example that the lower (upper) bound resolution threshold of the vision system is $h_l(h_u)$. Therefore, a measurement δ of an object or tool position in the 3-D Cartesian space is related to its true position δ_T with the following equation defined by Brooks [39]:

$$\delta(1 - h_l) \leq \delta_T \leq \delta(1 + h_u) \tag{4.4.1.51}$$

The above expression may be written analytically:

$$\delta_x(1 - h_l) \leq \delta_{Tx} \leq \delta_x(1 + h_u) \tag{4.4.1.52}$$

$$\delta_y(1 - h_l) \leq \delta_{Ty} \leq \delta_y(1 + h_u) \tag{4.4.1.53}$$

$$\delta_z(1 - h_l) \leq \delta_{Tz} \leq \delta_z(1 + h_u) \tag{4.4.1.54}$$

$$\delta_x^2 + \delta_y^2 + \delta_z^2 = \delta^2 \tag{4.4.1.55}$$

$$\delta_{Tx}^2 + \delta_{Ty}^2 + \delta_{Tz}^2 = \delta_T^2 \tag{4.4.1.56}$$

The true position of an object or tool is somewhere in the interval $[\delta(1 - h_l),\ \delta(1 + h_u)]$. The position of an object or tool is measured with an accuracy, h. Thus, there are:

$$\nu = \delta(h_u + h_l)/h \tag{4.4.1.57}$$

possible outcomes of a specific measurement δ. Assign a probability distribution function $p(\delta)$ to the possible outcomes of the measurement. The cost function of this operation is given in terms of the entropy function related to the measurement uncertainty. Considering a uniform distribution of the possible outcomes, the cost function is:

$$J_2 = H(p(1/\nu)) \qquad (4.4.1.58)$$

This distribution may not be necessarily a uniform distribution (results in maximum uncertainty). It may also be a singular distribution (results in minimum uncertainty), or any other justifiable distribution. Its specific form depends on the accuracy and resolution of the vision system and the distribution of objects and tools in the workspace environment.

The on line feedback information communicated from the execution devices to the Vision System Coordinator is:

1. Images of the objects and tools in the workspace environment, and,
2. Information about the location of each object or tool in terms of a (6x1) position and orientation vector $r = (x, y, z, \alpha, \beta, \gamma)^T$.

The feedback information is interpreted by the Vision System Coordinator and is used to calculate the accrued cost function related to its overall operation:

$$J_V = J_1 + J_2 \qquad (4.4.1.59)$$

given in terms of the associated entropies, where J_1 and J_2 are the two individual costs associated with the operations of the execution devices.

This feedback information and the individual costs and accrued cost are stored in the short term memory of the coordinator, which is a part of the overall memory of the dispatcher at the coordination level. They are used to calculate the overall accrued cost of the coordination level after the execution of the requested job.

The Functions of the Sensor System Coordinator

The purpose of the Sensor System Coordinator is to provide information about the real world environment of the Intelligent Robotic

System. It is defined in a broad way as perception. Quoting [26], *it refers to a mapping from the product space of relevant physical properties into the information space required for the execution of a given task.* In other words the Sensor System Coordinator establishes a mapping/correspondence between the workspace environment and the system model, monitors and maintains this correspondence in order for the system to function properly.

A variety of criteria can be considered when attempting to define and classify the functions of a sensory system: external sensing, internal sensing and other classification criteria which address the nature of the physical properties sensed. Specific hardware is necessary to interpret this sensory information. It is beyond the scope of this monograph to provide techniques for real time interpretation and use of sensory information. For our purposes it is sufficient to consider that the only function of the Sensor System Coordinator is to provide information about the force/torque required to be applied to a robot arm's gripper whenever the gripper has to grasp an object or tool. Details about this function are given in the corresponding section of the Gripper Sensor Coordinator. Therefore, the associated cost with the Sensor System Coordinator is embedded in the Gripper System Coordinator cost.

The Functions of the Motion System Coordinator

The purpose of the Motion System Coordinator is to choose the appropriate path(s) for the robot arm(s), select the best dynamic model (if more than one available) for the specific movements and assign end conditions and penalty functions associated with these motions. The end conditions are assigned based on each particular execution of a specific task and related feedback from the lower level. The purpose of the penalty functions is to increase the accuracy in the execution of these specific tasks. They are assigned in terms of weighting functions imposed by the specific mechanical actions. The execution devices associated with the Motion System Coordinator are the manipulators and their dedicated controllers. Figure 4.7 shows how one manipulator (there may be more than one) is connected and controlled by the Motion System Coordinator. The upper three degrees-of-freedom are used for the positioning of the hand and the lower three for the orientation of the wrist [117]. When the Motion System Coordinator is activated it issues a number of different commands to its associated manipulator(s):

1. Command of converting the visual data provided by the (6x1) vector $r = (x, y, z, \alpha, \beta, \gamma)^T$ to joint arm coordinates $q = (q_1, q_2, q_3, q_4, q_5, q_6)^T$ for a six degree-of-freedom manipulator.
2. Command of accessing the trajectory file in order to choose one trajectory for a particular motion. The trajectory is chosen based on the initial location of the robot arm(s), the location of object or tool that will be reached and the location of (possible) obstacles present in the workspace environment.
3. Command of choosing the dynamics model for the motions. The choice of the model depends on the chosen trajectory, speed and load conditions.
4. Command of choosing a particular control algorithm out of the available ones. The choice of the particular algorithm is based on the path (trajectory) chosen, the individual joint motions and the timing constraints. Controllability and stability of the control algorithms are necessary, and,
5. Command of assigning the cost functions, end conditions and penalty functions to the formulated control problem.

The research work in this area has already been completed [117, 234, 239]. The cost function associated with the Motion System Coordinator is given in terms of an appropriate performance index defined in classical control theory as [19]:

$$J_{MC} = J\{V(q(t), u(t), t)\} \tag{4.4.3.60}$$

where $V(q(t), u(t), t)$ is a function of the generalized arm coordinates (states) $q(t)$, control action $u(t)$ and time t. For end point control and adaptive control, the above performance index may be expressed as an entropy function by defining [195]:

$$H(x_0, u, p(u)) = E\{V(x_0, u, t)\} \tag{4.4.3.61}$$

and following the procedure described in chapter two. The on line feedback information communicated to the Motion System Coordinator from the associated execution devices (from the execution level) is a (6x1) vector of the final (end) position of the arm given in terms of the arm coordinates: $q_f = \left(q_{1f}, q_{2f}, q_{3f}, q_{4f}, q_{5f}, q_{6f}\right)^T$. This information is interpreted by the Motion System Coordinator and is used to calculate the positional and orientational errors of the end effector of the manipulator and also decide about adjustments to the final position and orientation of the manipulator.

The Functions of the Gripper System Coordinator

The purpose of the gripper coordinator is to pick up an object or tool from one place and move it to another safely. In order to do that, a command is sent by the Gripper System Coordinator to its associated execution device (the gripper at the execution level) shown in Figure 4.7 to grasp the object or tool and keep it during the whole motion. On line feedback information from the sensory and vision system provides information about the object's or tool's size and shape position and orientation as previously explained. This information is interpreted by the gripper in order to calculate the necessary opening and force/torque to grasp and hold the object or tool. The cost function associated with this operation is a *reward-punishment* type of performance index which represents the probability of a successful gripper operation, given in terms of an entropy function:

$$J_G = H(p(w)) \tag{4.4.4.62}$$

where $p(w)$ represents the probability of success.

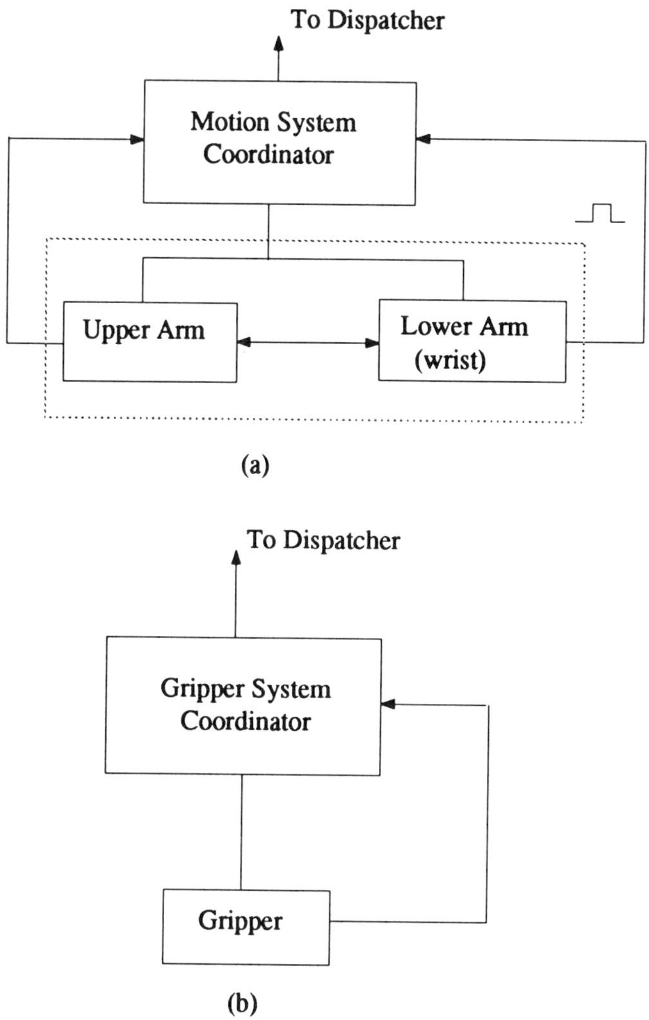

Figure 4.7 The Motion, Gripper System Coordinators and Execution Devices

The overall accrued cost of the coordination level associated with the actual execution of the requested job is given as the sum of the accrued costs of the individual coordinators:

$$J_c = J_V + J_{MC} + J_G \tag{4.4.4.63}$$

expressed in terms of the corresponding entropies. This cost is communicated to the organization level after the execution of the requested job and is used to upgrade the information stored in the long term memory of the organizer.

After the derivation of the individual functions of the different coordinators and the description of the on line feedback information from the execution to the coordination level, the algorithm of the coordination level follows.

The Algorithm of the Coordination Level

The steps of the coordination level algorithm are as follows:

1. The dispatcher scans the buffer with the stored plan and associates each coordinator with its corresponding execution devices from the execution level.
2. Formulate the control problem to be executed by the execution level. This involves activation of the execution devices and execution of specific tasks when their corresponding coordinator is accessed. In detail, formulation of the control problem includes:

   a. Collect information about objects and tools present in the workspace environment. Identify the proper objects and tools necessary to complete the requested job.
   b. Collect information about the initial and final location of the necessary objects and tools. Formulate the cartesian vector of the initial and final positions:

$$r_{in} = \left(r_x, r_y, r_z, \alpha, \beta, \gamma\right)^T \tag{4.4.5.64}$$

$$r_f = \left(r_{x'}, r_{y'}, r_{z'}, \alpha', \beta', \gamma'\right)^T \tag{4.4.5.65}$$

c. Convert the above information (when necessary) into the joint coordinates space:

$$r_{in} \rightarrow q_{in} = (q_1, q_2, q_3, q_4, q_5, q_6)^T \tag{4.4.5.66}$$

$$r_f \rightarrow q_f = (q_{1'}, q_{2'}, q_{3'}, q_{4'}, q_{5'}, q_{6'})^T \tag{4.4.5.67}$$

d. Based on the information from the vision system choose from the trajectory file $\Gamma(tr(t))$ the trajectory(ies) that will be used for the different motions.

e. Choose the manipulator(s) dynamics models (diagonal, diagonal dominant, complete), from the set Δ_m of the available ones.

f. Choose a particular control algorithm from the set of the available ones based on the kind of trajectories chosen (smooth, sharp) and the different motions (in favor of, or against the gravity, loaded, unloaded).

g. Choose a performance index, based on the problem constraints (i.e., minimum time, end point, adaptive, etc.)

3. Evaluate the execution level. This involves calculation of the accrued cost functions associated with the plan execution, based on the individual cost functions of the execution level.
4. Communicate selective feedback information received from the lower level to the organizer after the execution of the requested job.
5. Upgrade the information stored in the coordination level short term memory (in the dispatcher).

Probabilities are assigned to each trajectory, dynamics model and control algorithm. These probabilities are upgraded after the execution of the formulated plan with the aid of learning algorithms. The case studies in chapter six clarifies this approach.

4.5 The Execution Level

The execution level is composed of a number of execution devices related to the coordinators of the coordination level. Its main objective is to execute the specific tasks issued by the different coordinators as accurate as possible. Individual cost functions are assigned to each operation of the execution devices the form of which depends on the specific application problem. This requires system theoretic methods suitably implemented to control the hardware of the execution level. Many researchers have attempted to do that by linking System Theory with Information Theory. Their results did not find any important application because of lack of physical interpretation.

Saridis [195] has established entropy measures discussed in chapter 2 that are equivalent to the performance criteria of the optimal control problem while providing a physical meaning to the latter. This is done by expressing the problem of the control system design probabilistically and assigning a distribution function representing the uncertainty of selection of the optimal solution over the space of admissible controls. When the distribution satisfies Jaynes' maximum entropy criterion [93, 94], the performance criterion of the control problem is associated with the entropy of selecting a certain control. Minimization of the differential entropy which is equivalent to the average performance of the system yields the optimal control solution. The adaptive control formulation of the problem is derived as a special case of its optimal formulation. A summary of the overall system model is now presented for clarification purposes.

4.6 Summary of the Mathematical Model

When a user command is issued and sent to the Intelligent Robotic System, it is received and classified by the classifier to yield the compiled input command to the organization level. Then, the organization level formulates complete and compatible plans to execute the user requested job. This is done by associating the compiled input command with a set of pertinent activities (strings of binary valued random variables) and by organizing the activities (strings) in such a way (planning) to yield complete and compatible plans.

Two approaches may be followed depending on the amount of existing knowledge related to the requested job: a probabilistic approach or an expert system approach. The best plan generated (based on some criteria) is transferred to the coordination level. The coordination level decides about the best possible way to execute this particular plan. It issues specific commands to the execution level which is composed of a number of execution devices associated with the coordinators at the coordination level.

The interaction between the three levels is represented in terms of on line (real time) and off line feedback information. The on line feedback information is communicated from the execution to the coordination level during the execution of the requested job. The off line feedback information is communicated from the coordination to the organization level after the execution of the job and is used to evaluate and upgrade the information stored in the long term memory of the organization level (functions of feedback and long term memory exchange).

4.7 Integrated Control and Diagnostics in Intelligent Robotic Systems

Every mechanical system, hierarchical or not, does not always function properly. System operation is subject to soft or hard failures which degrade performance and may lead to major damages.

The models presented thus far, assume implicitly that the Intelligent Robotic System either functions properly or, should a failure occur, it is accommodated (somehow) by the sensory system.

However, one very important aspect of Intelligent Robotic Systems is related to their ability to distinguish between plan failures due to poor planning strategies and plan failures due to actuator and/or sensor (local) failures. While the former requires complete replanning, the latter requires *detection, identification, isolation* and *accommodation* of local failures. This is the type of machine intelligence which may be called called *diagnostic intelligence*, a term coined in [154], directly related to the ability of the system to overcome soft failures.

Many types and kinds of (local) failures may occur within an Intelligent Robotic System. Two very important (local) failures are related to malfunction of the Vision System Coordinator and Motion System Coordinator. The former type has to do with failure to recognize an already known object within the workspace environment, miscalculation of its correct position and orientation, as well as failure to recognize an "alien" object which enters the workspace. The latter type is directly associated with the robot payload variation problem, i.e., the accidental dropping of a load while the robot manipulator is moving from one location to another.

Failures related to the vision system are more difficult to accommodate and they are related to current advances in the area of Computer Vision, and in particular, stereo and introspective vision and hybrid intensity and range sensors. Identification and accommodation of this type of failures is beyond the scope of this monograph and, therefore, are not treated here.

On the other hand, the robot payload variation problem is better defined and may be overcome. A specific methodology is presented in the sequel.

Accommodation of the Robot Payload Variation Problem

This section describes the problem of integrated control and failure detection of payload variations during the execution of a task by incorporating failure diagnostic methods based on the novel four-parameter control configuration, the theoretical basis of which was developed in [154] and its application to the simultaneous analysis and design of control and failure diagnostic systems was demonstrated in [153]. The four-parameter controller design may be separated into two independent problems; that of the controller/compensator design and that of the failure detection system. Each of these processes involves the selection of two parameters. However, the combined controller and failure detection system is a nontrivial combination of parameters from both the controller and failure detection systems. The four-parameter controller when applied to simultaneous diagnostic and control systems provides a generic setting that may be used to study fundamental trade offs present in any simultaneous control and diagnostic system.

Given the basic configuration of a diagnostic and control system, shown in Figure 4.8, the objective (not only for robotic systems but also for general systems) is to design a controller that provides good control performance from the reference input w' to the controlled system output y and utilize the additional controller output z' to provide a diagnostic capability. The diagnostic capability of the four-parameter controller systematically utilized for the detection (and accommodation) of payload variations, is done by modeling failures in the system as disturbance inputs n and then designing the controller so that these failures are reflected or identified at z'.

The formulation of the problem of detecting abrupt payload variations as a failure diagnostic problem utilizes the linearity under nominal conditions of the robotic manipulator computed torque control technique. Each joint is controlled independently. Joint coupling due to model parameter mismatch is treated as a plant output disturbance. This allows the four-parameter controller design to be carried out analytically for each joint. For an n degree of freedom manipulator, n of these controllers operating in parallel are designed. The goal of the diagnostic capability output is to detect with high reliability the dropping of the load mass as soon as possible after the dropping occurs. The control configuration shown in Figure 4.8 represents the most general interconnection of a given plant with a controller that preserves the distinction between the plant and the controller. Since we are interested in discussing generic controllers and not plants, the four-parameter control configuration is restricted to the one shown in

Figure 4.9. In this figure, for a generic plant, u represents the inputs that may be manipulated, w represents the inputs that may not be manipulated (disturbance inputs), y represents signals available for control, z represents signals to be controlled w' is the reference input and z' the diagnostic output. The exogenous signals n, n' are deviations from the ideal plant G (where ideal means nominal and noiseless). In order to distinguish between noise and failures, we consider that $n = f+\eta$ and $n'=f'+\eta'$and define f, f' as failures and η, η' as noise. Correspondingly, y, y' are regarded as ideal sensor and actuator outputs and u, u' as actual sensor and actuator outputs, respectively.

The action of the controller K is interpreted as a matrix, namely:

$$\begin{bmatrix} z' \\ y' \end{bmatrix} = \begin{bmatrix} K_{11} & K_{12} \\ K_{21} & K_{22} \end{bmatrix} \begin{bmatrix} w' \\ u' \end{bmatrix} \tag{4.7.1.68}$$

where K_{ij} are four matrix parameters describing the action of the controller.

The diagnostic objective is to follow or identify $\tilde{f}$ at z', where:

$$\tilde{f} = \begin{bmatrix} f \\ f' \end{bmatrix} \tag{4.7.1.69}$$

Implicit in this objective is the need to reject all other system inputs reflected at z' namely, the sensor and actuator noise η, η', the reference input w' and the plant disturbance input w. The identification of $\tilde{f}$ is also done in the face of plant model errors. Thus, the diagnostic objective may be summarized as follows: in the face of plant modeling errors, achieve closed loop stability, and also make $M_{\tilde{f}z'} - I, M_{n'z'}, M_{nz'}, M_{w'z'}, M_{wz'}$, 'small' in an appropriate sense. Here M_{xy} denotes the transfer function from signal x to signal y.

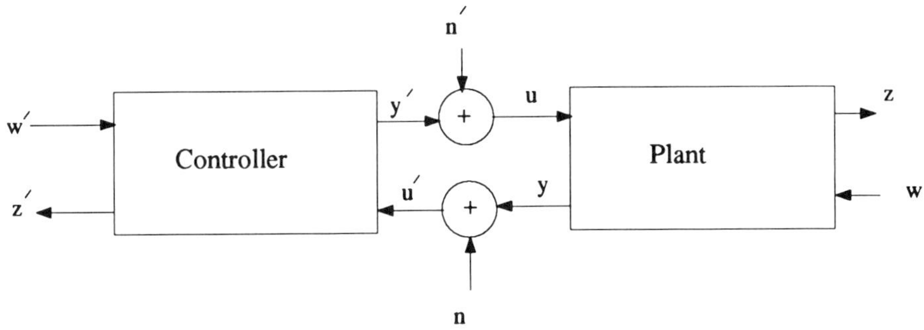

Figure 4.8 Generic Control System Configuration

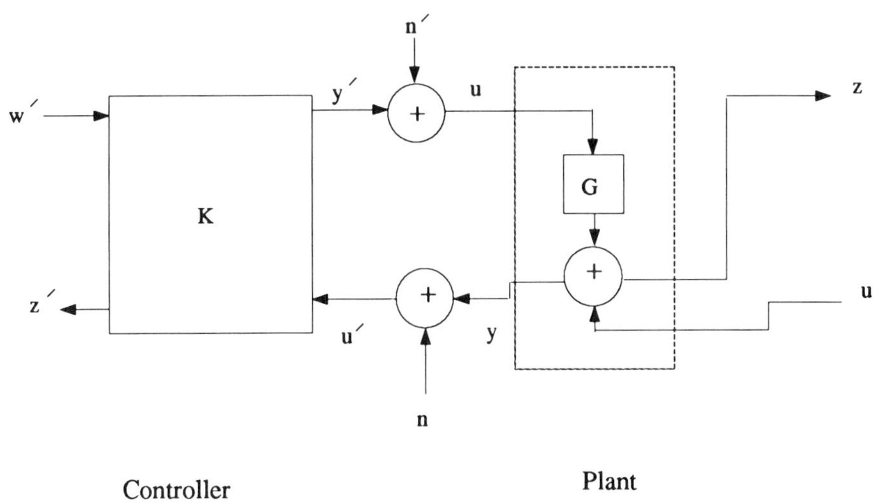

Figure 4.9 Generic Controller, Restricted Plant

The first requirement dictates that the failures, $\tilde{f}$, are identified in the diagnostic output signal z'. The next two, that the noise signals are attenuated in the diagnostic output, and the last two that the reference and disturbance signals are decoupled from the diagnostic output [153, 154].

The fundamental limitations of performing both diagnostic and control actions are addressed via a parameterization of all stabilizing four-parameter controllers for a given linear time invariant finite-dimensional plant G that lays bare all the input/output maps of the system in a particularly simple fashion. This restriction is made only for simplicity as other more general modeling descriptions may be utilized [154]. The parameterization of all stabilizing four-parameter controllers for a possibly unstable plant G is obtained through a doubly coprime fractional representation of G [246]:

$$G = ND^{-1} = \tilde{D}^{-1}\tilde{N} \tag{4.7.1.70}$$

$$\begin{bmatrix} V & U \\ -\tilde{N} & \tilde{D} \end{bmatrix} \begin{bmatrix} D & -\tilde{U} \\ N & \tilde{V} \end{bmatrix} = I \tag{4.7.1.71}$$

where $\tilde{U}, \tilde{V}, \tilde{N}, \tilde{D}, N, D, U, V$ are all stable matrices of appropriate dimensions. Here stable refers to the usual bounded input bounded output (BIBO) notion of stability for LTI finite-dimensional systems. It is also possible to incorporate certain performance issues in a more generalized notion of stability [153, 154, 217, 246]. Given a state space representation of G such a factorization can be easily computed and all stabilizing compensators can be explicitly parameterized as has been shown in [154] in terms of the above stable matrices and for some stable C with:

$$C = \begin{bmatrix} C_{11} & C_{12} \\ C_{21} & C_{22} \end{bmatrix} \tag{4.7.1.72}$$

with the dimension of C_{22} those of G transpose.

This parameterization may be interpreted in terms of an 'implicit internal model principle' (discussed at length in [154]) that allows a design separation to be seen for a simultaneous diagnostic and control system. The input output maps may be written in terms of both K_{ij} and the free parameters C_{ij} as explicitly derived in [153, 154].

The main advantages of the four-parameter controller are that any control system implemented simultaneously with a diagnostic system may be placed into this framework and that this framework allows the interaction between the control and diagnostic system to be clearly seen and quantified. The design freedom in using a diagnostic system with a feedback control system is addressed for the first time in the four-parameter control configuration. This general setting is proposed as a methodology to detect robot payload variations. An assumption made through the design is that modeling errors due to robot model mismatch are larger than any sensor or actuator noise. Of course, a very accurate design requires identification and calculation of the respective noise bounds (as has been done with other non robotic applications [153]).

As a case study, the four-parameter controller approach has been applied and tested by simulation on the first three degrees of freedom of a PUMA-560 robot manipulator. The PUMA manipulator is a high gear articulated manipulator and is a valid case study as shown in [239] and references therein. The four-parameter controller design may be also used to accommodate payload variations in other types of robots (SCARA, Direct Drive).

Concentrating in the PUMA manipulator, a four-parameter based PD computed torque controller has been designed for each of the higher three degrees of freedom of the arm, while the simulation studies have been performed by using trajectories achievable in real time and realistic system hardware constraints. The block diagram of the model of an ideally linearized robot arm controlled by the four parameter controller system is shown in Figure 4.10 for each degree of freedom, where w' represents the set of desired position, velocity and acceleration trajectories, w the disturbance (error) position and velocity trajectories, y the actual/measured position and velocity trajectories, and G the robot (plant) model. The PUMA model used is the same with the one used by Spong and Vidyasagar [217]. It was found that immediately after a load is accidentally dropped, the acceleration error behaves like a step function. Therefore, a filter may be designed for each joint to detect the occurence of the step. The four parameters involved in this design, were chosen based on the previous observation. A very detailed design of the four-parameter controller is given in [239]. The results have demonstrated that payload variations higher than 0.5 kg are identifiable in z' when they are accidentally dropped at several time instants, i.e., 1.5, 2.0, 2.5 seconds.

As far as integration of this mechanism with the rest of the system is concerned, the simplest way is to consider z' as part of the on line feedback information to the coordination level. No dedicated hardware is necessary to implement the four-parameter controller arround each joint. A binary threshold logic approach may be followed to distinguish between proper functioning of the manipulator, or malfunction. This topic is an open research issue of most importance.

4.8 Discussion

The mathematical model is sufficiently general to be applied to any type of Intelligent Machines and hierarchical systems. The limitations imposed on the operation of the organization level imply some simplifications within the algorithm of the organizer, and at the same time preserve the generality of the algorithm. Further modification of the operational procedures of the organization level may account for more or less design constraints. The advantage of the organization level algorithm is the bottom-up approach followed to formulate and organize plan scenarios. Plans are formulated from a set of primitives appropriately arranged for sequential execution.

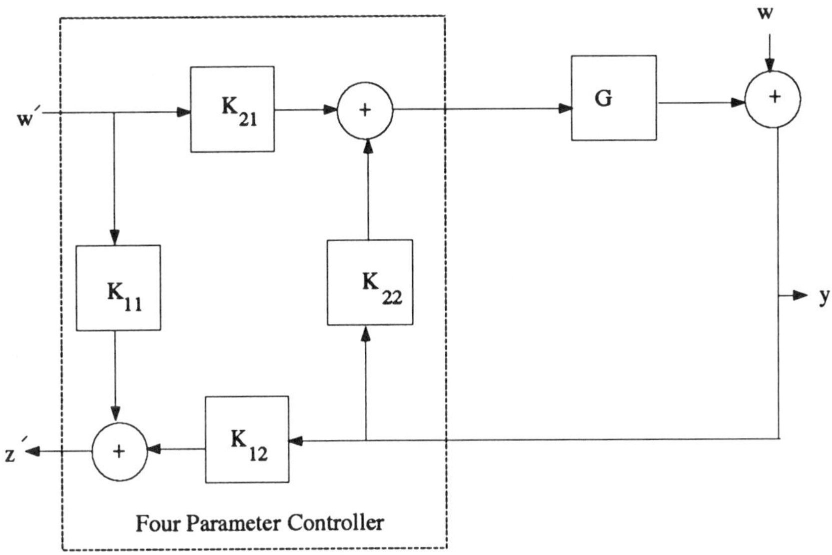

Figure 4.10 Block Diagram of a Model of an Ideally Linearized Robot Arm Controlled by a Four-parameter Controller

The function of the coordination level is described in terms of the functions of its different coordinators. The integrated control and diagnostics model accounts for unpredictable events during plan execution related to robot payload variations, but it does not consider a complete sensory system coordinator. It may be modified and extended to detect and accommodate several other potential failures.

Based on the last remark, it is essential to clarify and justify further the issues of *replanning, emergency replanning, local goals and global goal.* Replanning is generation of alternate plans when the already chosen plan is unsuccessful in executing/completing the user requested job. Replanning does not consider any kind or type of potential local failures, i.e., it is assumed that all system components function properly. Therefore, replanning occurs when the chosen plan fails as a whole.

Emergency replanning is defined as generation of alternate execution scenarios associated with a specific plan (the chosen plan) as a result of a local failure, i.e., the robot dropped a load, the vision system is unable to recognize an object.

In other words, replanning corresponds to generation of alternate plan scenarios from the organization level while emergency replanning corresponds to generation of alternate execution scenarios associated with a specific plan, to accommodate local failures.

Thus, it should be clear that accomplishment of the local goals of the organizer does not necessarily imply accomplishment of the global goal. This is justified by the fact that local goals are achieved off line while the global goal involves real time execution which may include some unpredictable situations.

4.9 Remarks

The mathematical modeling for Intelligent Robotic Systems presented in this chapter, may be implemented by hardware in digital computers. The specific hardware mechanisms are described in the next chapter. However, no specific hardware mechanism is proposed for the integrated control and diagnostics system.

Chapter 5
ARCHITECTURAL MODELS FOR INTELLIGENT ROBOTIC SYSTEMS

5.1 Introduction

The mathematical modeling for Intelligent Robotic Systems presented in the previous chapter, is suitable for implementation on a general purpose high level digital computer. In this implementation, linguistic elements (user commands) and mathematical algorithms are described in an appropriate machine code, and this code may be operated on by software and/or hardware elements such as lexical scanners, parsers, translators, registers and upgrading routines. The proposed system architecture partitions the overall system into subsystems and derives a computational module for each subsystem based on the principle of Increasing Precision with Decreasing Intelligence. It interconnects all subsystems into a tree like structure producing multi layer hierarchies.

As justified previously, hierarchical systems structured according to the principle of Increasing Precision with Decreasing Intelligence have inherent potential for top-down hierarchical knowledge (information) processing. Each level has a functional task to perform. The functional task of each level is performed in the best possible way based on accumulated information and related feedback from the lower levels; this is level optimization, not overall optimization. Upon completion of this functional task, a command is issued to the immediate lower level, and the functional task of the lower level is accomplished. With this structure, each level evaluates and controls the performance of the immediate lower one. The architectural model associated with top-down hierarchical knowledge (information) processing is suitable for the *decision phase* of Intelligent Robotic Systems. The decision phase involves formulating complete and compatible plans, deciding which one is the best to execute the requested job and how to execute it. Therefore, it may accept special simple architectures specifically designed to implement the hierarchically intelligent control algorithms.

Specific hardware units are built within each of the higher two levels for the *upgrade phase* of the system. This phase involves the upgrade of the individual and accrued costs, probability distribution functions and knowledge base associated with a particular plan. Upgrade follows plan execution (completion of the decision phase) and is performed in a bottom-up way: costs and probabilities associated with the lower level are upgraded first, followed by the ones with the higher level. A complete upgrade of the knowledge base is also performed.

The architectural models for the three interactive levels describe the hardware units including special memories to store permanent and temporary information, memories to store probabilities, costs and terminal sets, and other special hardware units for each level separately.

The architectural models presented in this chapter, for both of the decision and upgrade phases, are appropriate for the two modes of operation of such a system: the *training* mode and the *well trained* mode. The training mode of operation is defined as the mode in which the system explores and learns its capabilities and alternative actions given the user requested job. While in this mode, all related mathematical functions are significantly modifiable by the learning algorithms rewarding certain plans and penalizing others [65, 66, 76, 77, 132–134]. The well trained mode of operation follows the training mode and is defined as the mode in which the system is well trained and knows exactly the sequence of actions necessary to complete a user requested job [237]. The well trained mode explained here is suitable when there are not situations which might include unpredictable events. In order to account for unpredictable events, the coordination level algorithm (and therefore its corresponding architecture) must be modified to account for emergency situations. To be more specific, a complete hardware architecture should be designed to interface the four-parameter based controller with the rest of the system.

The decision phase architectures associated with the two models of the organization level are presented first, followed by the models for the coordination and execution levels. The specific hardware units for the upgrade phase (on line and off line feedback mechanisms) are then described.

5.2 The Architectural Model for the Organization Level

The organization level performs general knowledge (information) processing tasks in association with a long term memory. The architectural model of the organizer accommodates fast and reliable operation for each individual function of the organization level. The mechanisms of the memory exchange function associated with the machine reasoning, planning and decision making functions are analyzed in detail. When describing the architectural units, it has been assumed that every string is compatible, therefore, the necessary memory required to store all activities, ordered activities, etc., indicates the worst case scenario. In reality, the storage requirements are much smaller.

The Probabilistic Based Architectural Model

The model for the machine reasoning function (passive) is shown in Figure 5.1. The input to the machine reasoning model is the (classified) compiled input command u_j and the corresponding output, Z_j^R, the set of the (maximum)$(2^N - 1)$ pertinent activities A_{jm} (strings X_{jm}) with the corresponding addresses which store the activity probability distribution functions $P(X_{jm}/u_j)$. The reasoning block *RB* contains a maximum of $(2^N - 1)$ strings of binary valued random variables $X_{jm} = (x_1, x_2, ..., x_{N-L}, ..., x_N)_m$, $m = 1, 2, ..., (2^N - 1)$, stored at a particular order, which represent the activities (A_{jm}) associated with *any* compiled input command. The corresponding addresses which store the probability distribution functions related to these activities (strings) are transferred from the memory D^R, a part of the long term memory of the organizer. The memory D^R consists of M different memory blocks $D_1, D_2, ..., D_M$. One memory block, D_j, is associated with each compiled input command u_j. Each block D_j, contains the specific addresses which store the elements of the probability vector S_j that corresponds to u_j. Once the compiled input command u_j has been recognized, realization of the switch s_1 activates (enables) the memory block D_j. Transfer of the data contained in D_j is accomplished via the realization of the switch s_2. The switches are coupled with each other. The contents of the *RB* are transferred to the right most positions of the *PRB* (probabilistic reasoning block) while the corresponding addresses occupy the left most positions. The information stored in D^R is not modifiable by, or during the machine reasoning function. Therefore, D^R is considered as *permanent* memory whose values *do not change during an iteration cycle*, i.e., from the time the user has requested a job until its actual execution. The values of the probability distribution functions associated with the set of pertinent activities stored in the addresses are upgraded *only after* the completion of the requested job through a specific hardware unit described in the sequel.

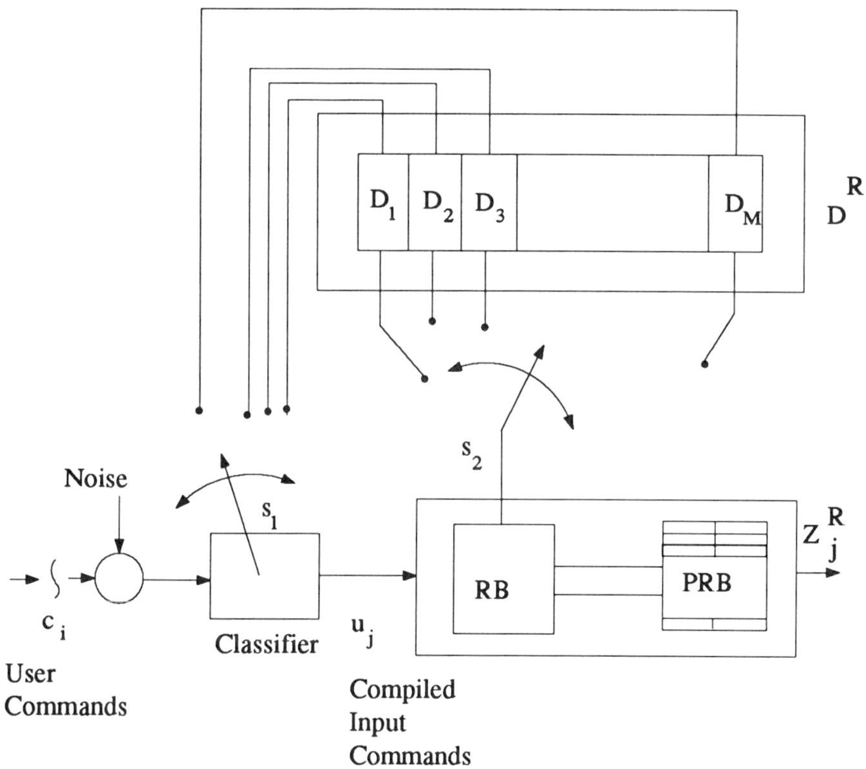

Figure 5.1 The Model for the Machine Reasoning Function

The model for the machine planning function is more complicated. It is shown in Figure 5.2. The input to the planning model is Z_j^R, or equivalently, the corresponding activated memory block. The output from the machine planning model is $Z_j^P = Z_{jmv}$, the set of all complete and compatible plans capable of completing a requested job. This set is a subset of the set of all compatible augmented ordered strings of primitive events formulated during the machine planning function. All compatible ordered activities (strings) (valid permutations of primitive events) are stored in the first planning box, *PB1*. Every compatible augmented ordered activity (string) (which includes the valid sequences of repetitive events) is stored in the second planning box, *PB2*. The compatibility tests are performed within the boxes *CPT1* and *CPT2*. The specific hardware unit for the compatibility tests is shown in Figure 5.2(b). The memory D_T, also a part of the long term memory of the organizer, is divided into four sub blocks, D_{T1}, D_{T2}, D_{T3} and D_{T4} : D_{T1} contains all incompatible pairs of the form (non repetitive primitive, non repetitive primitive), D_{T2} of the form (non repetitive primitive, repetitive primitive), D_{T3} of the form (repetitive primitive, non repetitive primitive) and D_{T4} of the form (repetitive primitive, repetitive primitive). Each ordered activity (augmented ordered activity) is transferred to the register $R \quad (R')$. At the beginning of the compatibility test the pointer $PT \quad (PT')$ is at the left most position of $R \quad (R')$. It transfers every pair of primitives to the two position register $R1 \ (R1')$, which scans D_{T1} (D_{T1}, D_{T2}, D_{T3} and D_{T4}) to check if the stored pair is incompatible. If yes, the whole ordered activity (augmented ordered activity) is rejected. If not, the left most primitive event in $R1$ $(R1')$ is discarded, the right most moves one position to the left and a new primitive event occupies the empty position. The test continues until the pointer has reached the last two primitive events of the activity. Therefore, the number of compatible activities (strings) is significantly reduced. The precedence constraint is also implemented in this way.

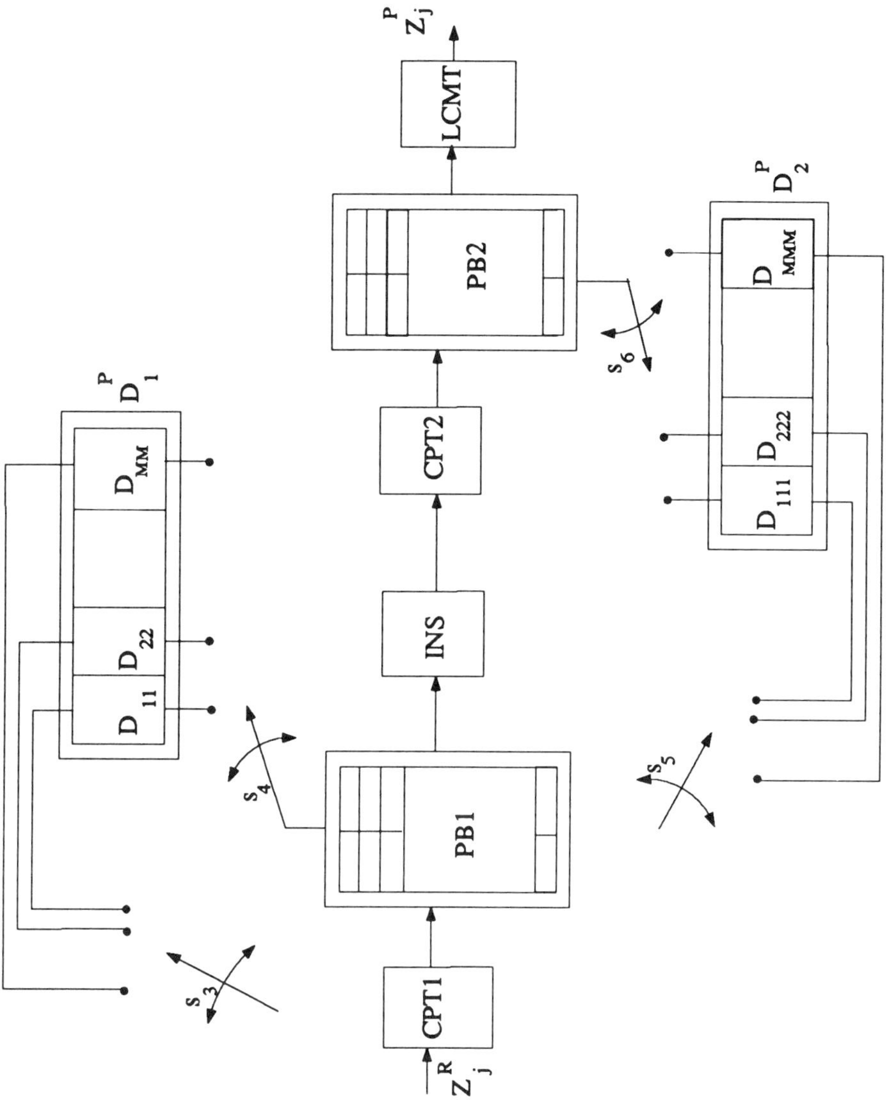

Figure 5.2 The Model for the Machine Planning Function

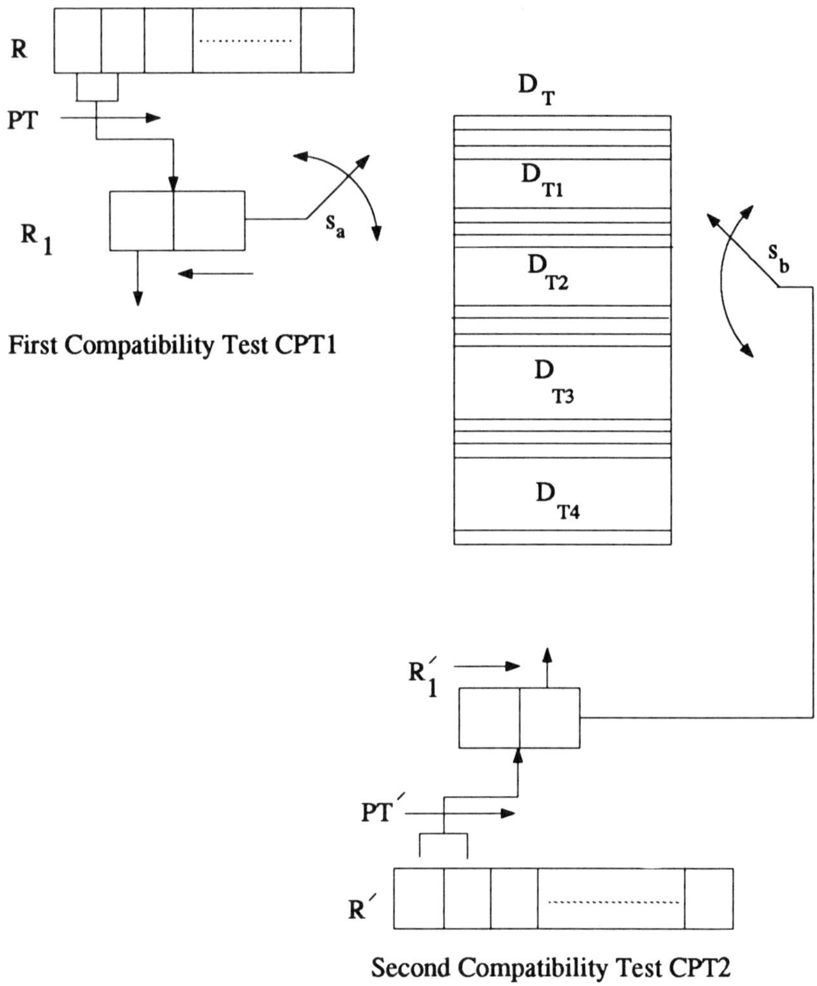

Figure 5.2 (Continued)

The corresponding addresses of the mask matrices with probabilities $p(M_{jmr}/u_j)$ are transferred from the memory D_1^P via the realization of the coupled switches s_3 and s_4 and are multiplied by the probability distribution functions $P(X_{jm}/u_j)$ to yield the probability distribution functions of the compatible ordered activities (strings) Y_{jmr}. The switch s_3 activates the corresponding memory block D_{jj} (once Z_j^R has been recognized), while the switch s_4 permits the transfer of data. The box *PB1* now contains the compatible ordered activities (strings) with their corresponding addresses containing the probability distribution functions $P(Y_{jmr}/u_j)$. The insertion of the valid sequences of repetitive primitive events is performed in the box *INS* as previously explained while the compatible augmented ordered activities are stored in the second planning box, $PB2$. The corresponding probabilities are transferred from D_2^P. Activation and transfer of data from the appropriate memory block D_{jjj} is accomplished via the realization of the two coupled switches s_5 and s_6 in a way similar to the one described before. The box $PB2$ contains now the augmented compatible ordered activities (strings) with their corresponding addresses containing the pertinent probability distribution functions. The information stored in D_1^P and D_2^P is not modifiable by, or during the machine planning function. This information is considered permanent within an iteration cycle, too. The values of the probability distribution functions stored in the addresses are upgraded after the completion of the requested job. The output from the machine planning function Z_j^P is the set of all complete and compatible ordered activities (strings) that may execute the requested job. The completeness test is performed within *LCMT*. This test accepts every meaningful syntactically correct plan as explained through the operational procedures. Every compatible augmented ordered activity (string) is checked. If a compatible augmented ordered activity (string) does not begin with a repetitive primitive event and end with a non repetitive one and does not follow the limitations imposed by the operational procedures, is rejected as incomplete. Hence, *every complete plan is compatible but every compatible plan is not necessarily complete.*

The model for the machine decision making function is shown in Figure 5.3. All complete and compatible plans are stored in *MDMB* (machine decision making box) and checked by pairs to find the most probable one. The most probable plan is stored in RR. If during the check, a complete plan with higher probability than the one already stored in RR is found, it is transferred to RR while the already stored one is discarded. Once the check is over, the contents of RR indicate the most probable complete and compatible plan to execute the requested job.

Every complete and compatible plan is also stored in a particular part of the long term memory of the organizer, D_{ss}. This memory contains every complete and compatible plan related to every compiled input command as shown in Figure 5.3. The idea of this particular memory is very important: It represents the situation of a well trained system (under the assumption that no unpredictable events may occur). An Intelligent Robotic System which has reached this mode of operation associates immediately after the recognition of the compiled input command the most probable of the plans (more than one available) stored in D_{ss}, without going through every single individual function as further demonstrated in the next section.

It is rather difficult to define when a system has reached its well trained mode. Such a definition depends on the application problem. Although this is the case, some indication may be obtained by looking at the evolution of the probability distribution functions; i.e., the probabilities of certain plans are very high (approach asymptotically one) and they do not change significantly after a certain number of iterations while those of other plans are almost zero. When this occurs, it is reasonable to deduct that the system has reached its well trained mode.

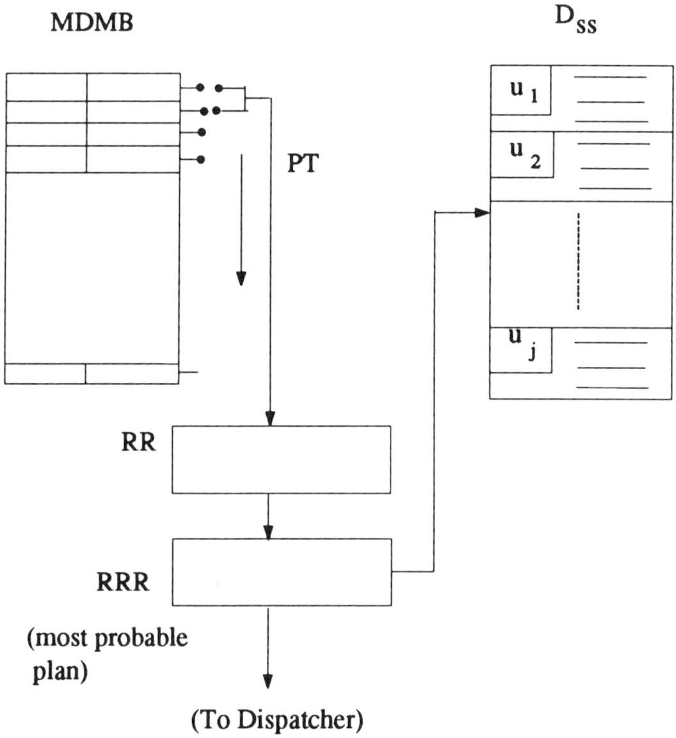

Figure 5.3 The Model for the Machine Decision Making Function

The Expert System Cell Based Architectural Model

The architectural model for the expert system cell configuration of the organization level follows, in principle, steps similar to the probabilistic model. This model represents a valid hardware alternate, suitable to the machine intelligence functions.

The classification technique, which is the same for both models, may be implemented in hardware using an abstractor, a NOR gate, a few latches and registers as shown in Figure 5.4. The user command gets filtered through the abstractor and is classified into a specific command type which is then simultaneously compared to all existing command types residing in the knowledge base. When determined whether or not the user command is of a new command type, a flag is set accordingly and relevant registers are loaded for the next processing stage.

Figure 5.5 shows the hardware implementation of the forward chaining method. It starts with the classified command type and command processing from the output of the classification stage. This method produces the task sequence which will carry out the user command that is ultimately executed by the hardware. The forward chaining method uses many different data structures of the knowledge base which need to be stored in fast static RAMs for optimal performance. Other hardware necessary to implement the forward chaining technique includes a counter, multiplexers, comparators, latches, registers, and logic gates. The basic structure of the sequence that is generated by the forward chaining method must exist in the knowledge base. Once a task plan is composed, it is simultaneously compared to all known previously failed sequences for that particular command, in the failed plan table. Should there be a match, this plan is automatically rejected due to previous unsuccessful trials of the plan. Finally, when the target sequence passes these tests, it is forwarded to the coordination level for further processing.

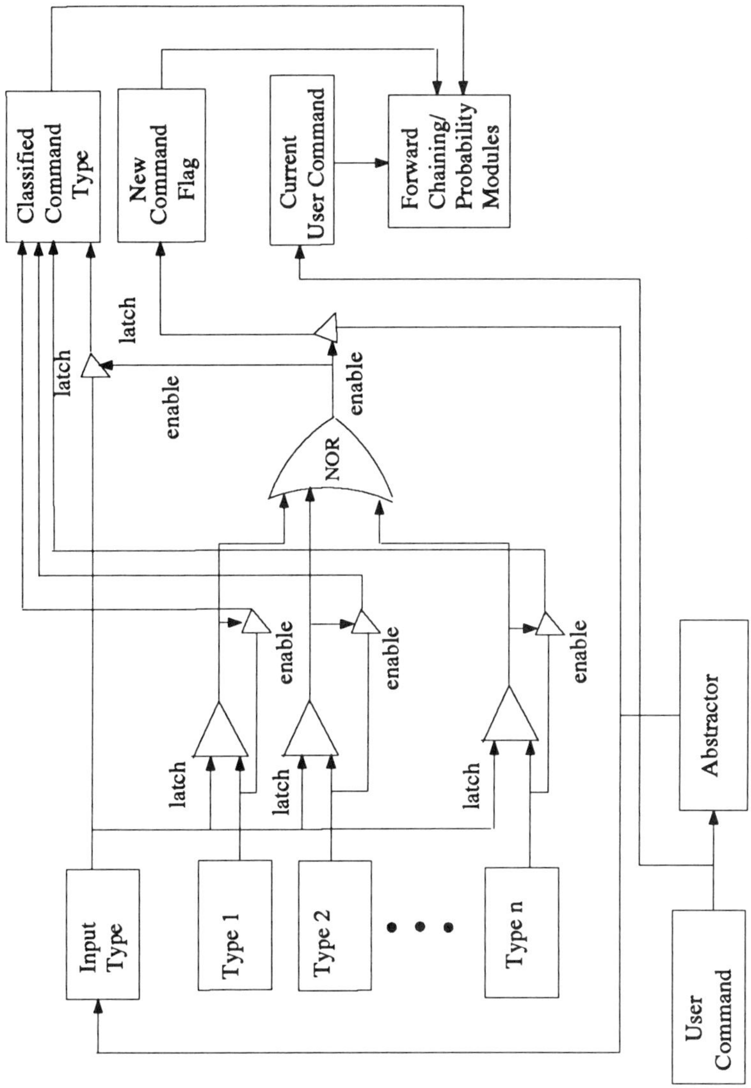

Figure 5.4 Hardware Implementation of the Classifier Model

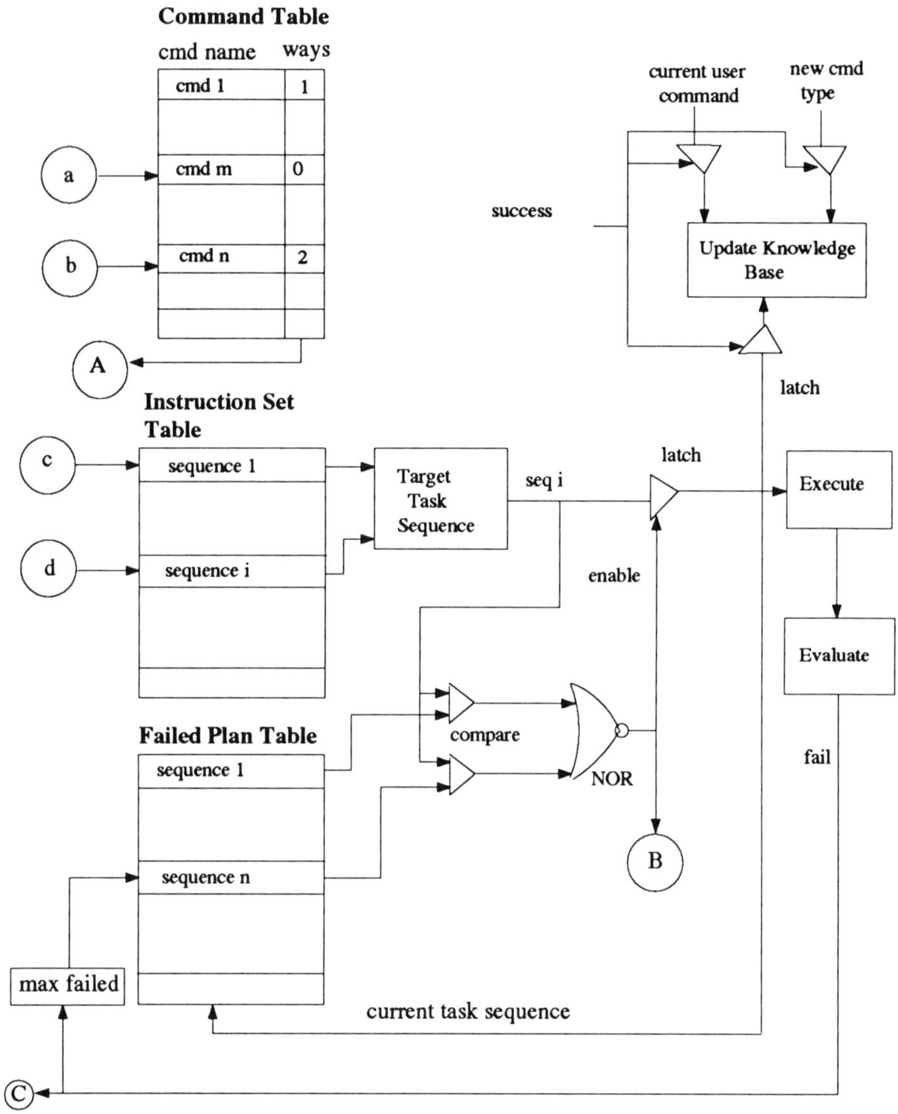

Figure 5.5 Hardware for the Forward Chaining Method

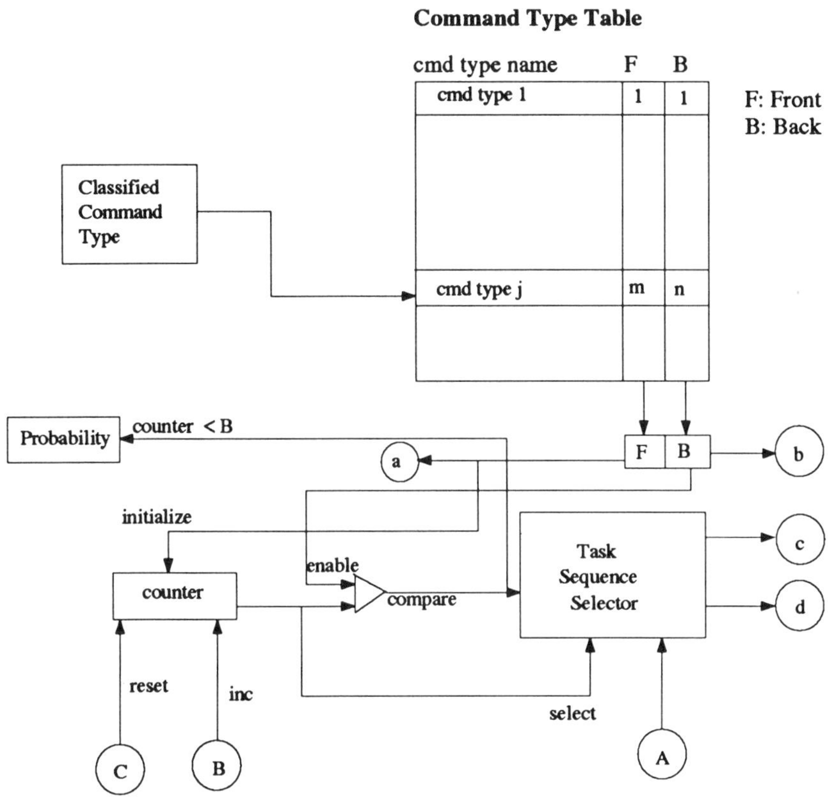

Figure 5.5 (Continued)

The strategy for the probabilistic part of this approach is similar to that of forward chaining, with the difference being that the basic sequences do not have to exist at all in the knowledge base. The main functional unit of the probabilistic method is the permutator with biasing and feedback as shown in Figures 5.6 and 5.7. Each basic task function, when added to the knowledge base, is determined at the outset either to be repetitive or non repetitive for a particular command or command type. If a basic task function is non repetitive, then that function can only be used by certain command types in the knowledge base; otherwise, all commands of all command types are allowed to incorporate that function in its task sequence. The permutator utilizes this information to generate task plans. From these resulting plans, the one with the highest probability is selected. All plans that have failed for that command will again be recorded in the failed plan table so that no previously failed task sequence could be issued again.

5.3 The Architectural Model for the Coordination Level

One general model is proposed for the coordination level. This model is suitable for both of the organization level models.

The coordination level consists of different coordinators, with dedicated microprocessors for each one of them. Once the best task sequence (complete plan) generated and chosen by the organizer, is sent to the coordination level, all the necessary details are provided, so that the chosen plan is successfully executed. Parallel tasks are performed where possible to achieve minimum time performance. An alternative block diagram for the coordination level is shown in Figure 5.8. The architecture is also inherently horizontal in that data is shared between the coordinators through the dispatcher. However, this does not violate the overall tree like structure. In order to execute tasks simultaneously, the Parallel Processing Task Scheduler (PPTS) assigns time stamps to each process and synchronizes these operations with the Parallel Processing Job Coordinator (PPJC).

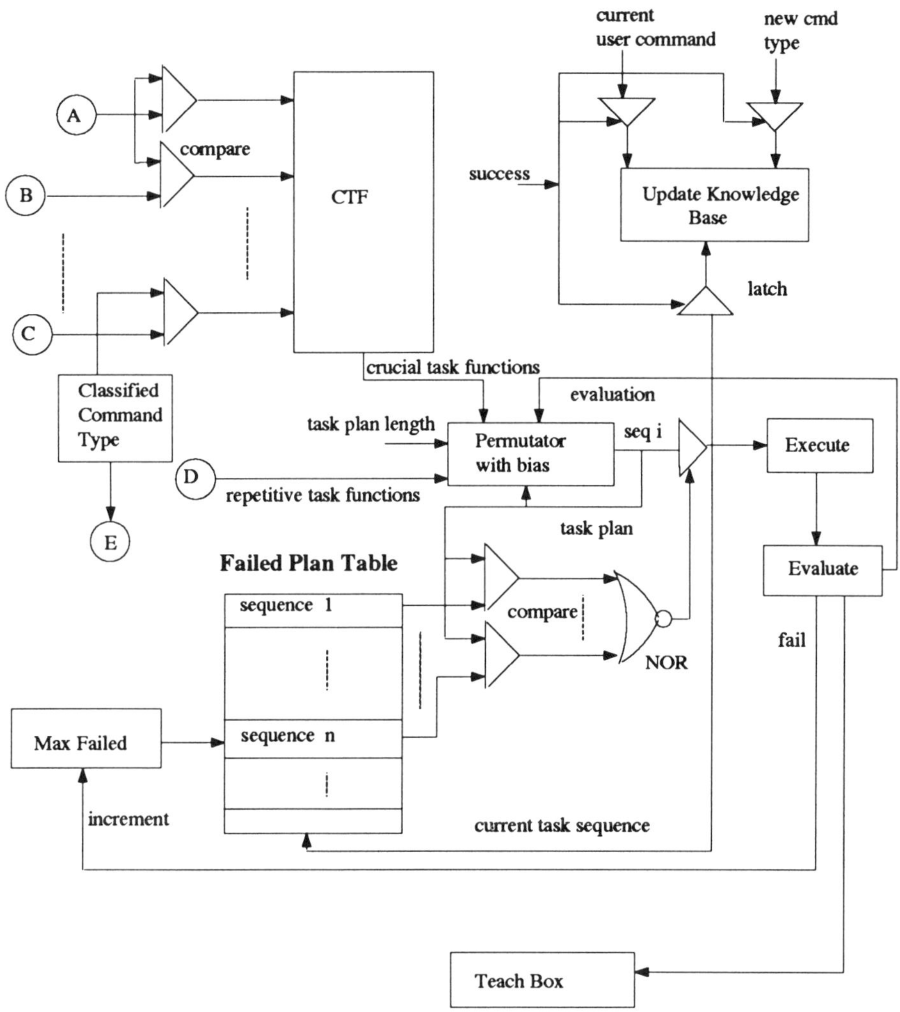

Figure 5.6 Hardware for the Probabilistic Method

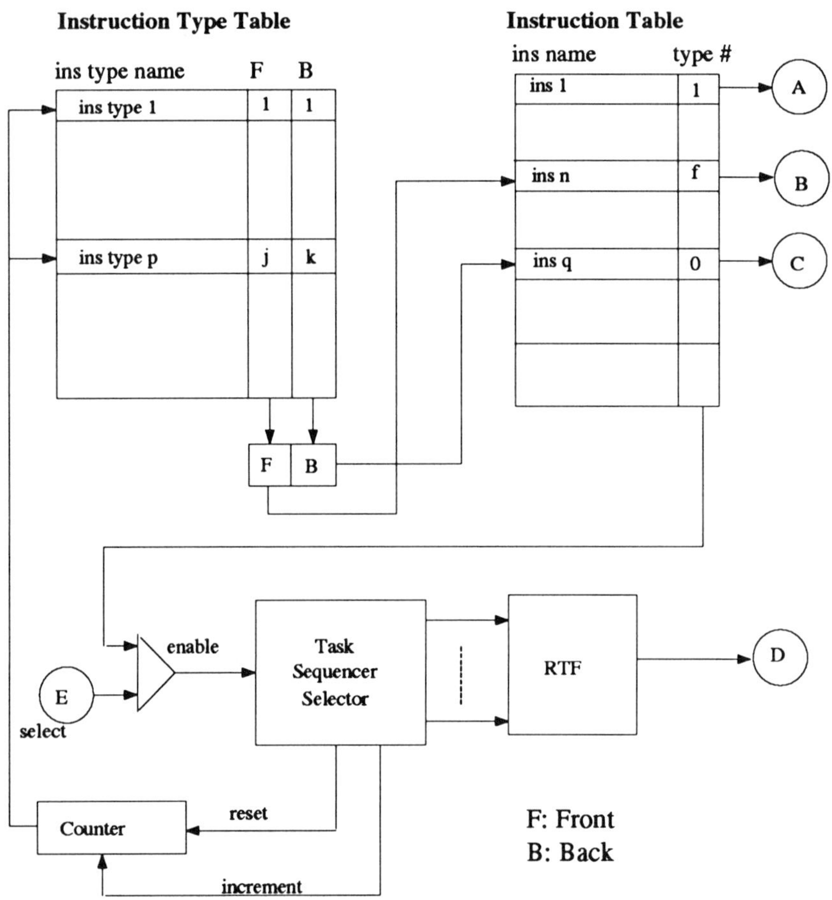

Figure 5.6 (Continued)

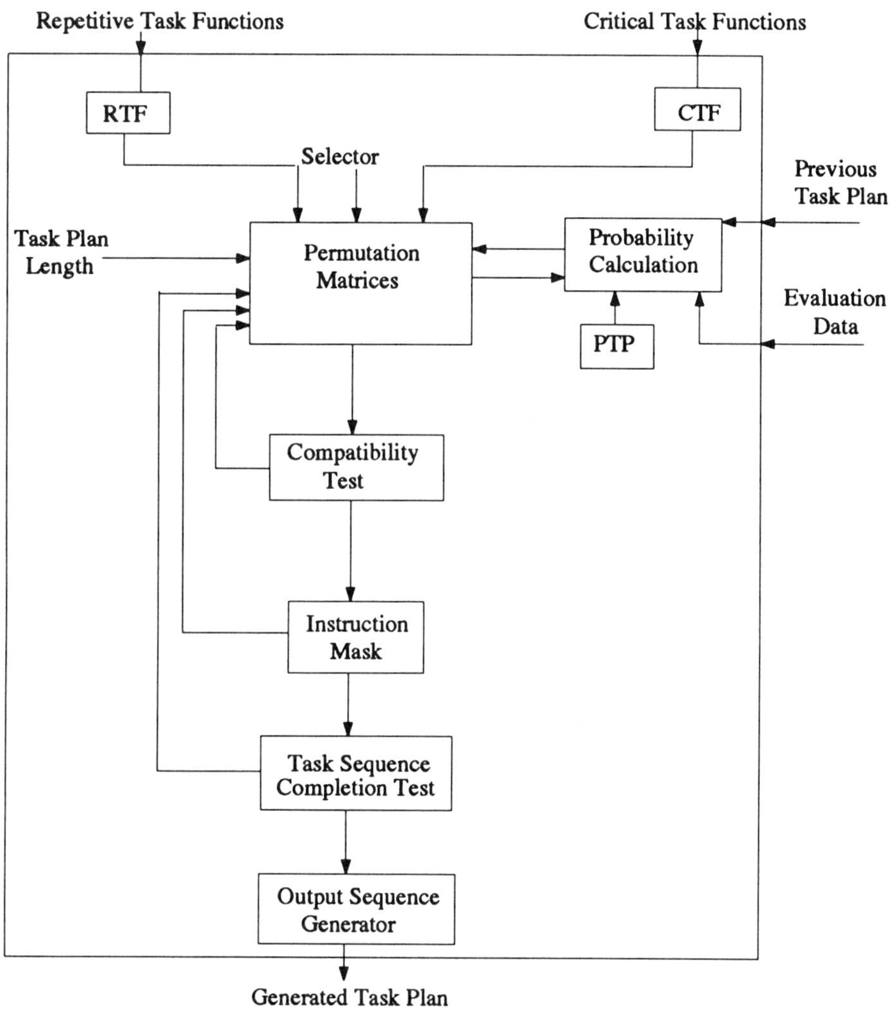

Figure 5.7 Permutator Hardware with Biasing and Feedback

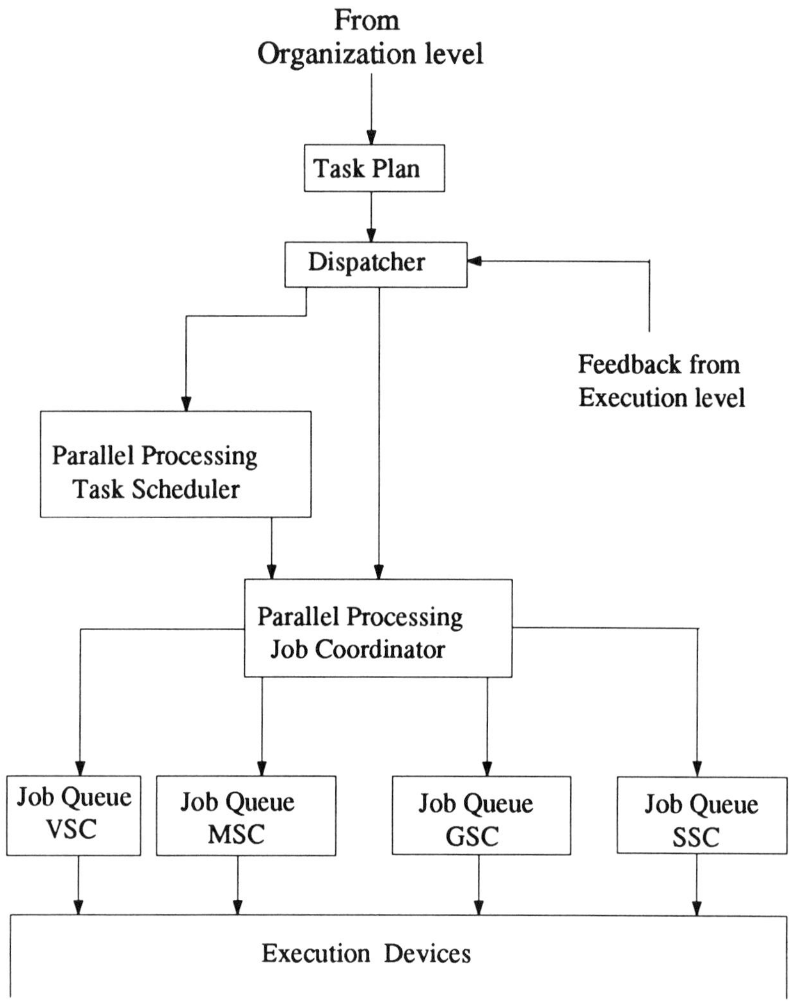

Figure 5.8 Alternative Block Diagram for the Coordination Level

The purpose of the PPJC is to control the flow of task execution based upon the time stamp given to each of these subprocesses and the availability of the different execution devices. The assignment of a time stamp is based upon the sequence of task execution and the dependability of one task to the next. To determine whether two tasks in sequence could be executed in parallel, the PPTS searches through the Parallel Execution Table (PET) for that particular ordered pair. If the pair exists in this table, then these tasks are assigned with the same time stamp indicating that they could be executed simultaneously. An algorithm has been developed for time stamp assignments which is shown in Figure 5.9.

The main hardware for implementing a typical coordinator is given in Figure 5.10. Each of the special purpose microprocessors forms the link between the organization and execution levels through their I/O ports. These microprocessor based systems use local ROMs for storing routines which are necessary for controlling the execution devices, and also RAMs for storing temporary information.

To be more specific, consider Figure 5.11 for a better understanding of the coordination level functions. The complete and compatible most probable (final) plan, contains actually a sequence of addresses. Each address corresponds to a primitive event. When the final plan is loaded into the buffer, the BEVENT line is deactivated and the pointer PT1 scans the buffer once from left to right. With the aid of the qualifier QLFR it is known how many times each coordinator will be accessed, as well as, when it will be accessed. Thus, the associated execution devices are activated when their coordinator is accessed via the qualifier QLFR. After the first scanning, the pointer PT1 returns to the left most position. The BEVENT line is deactivated and the qualifier is used to activate another coordinator as the pointer moves from left to right. Upon completion of its operation a BENT signal is sent to the qualifier and the buffer and the pointer moves one position to the right.

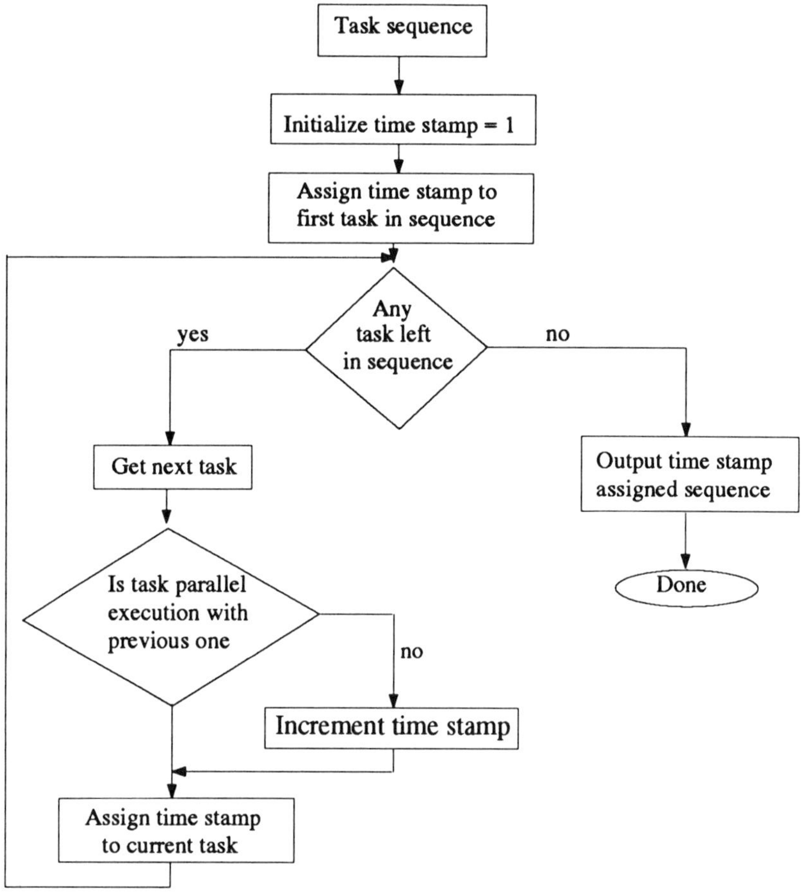

Figure 5.9 Time Stamp Assignment for the Parallel Task Execution Algorithm

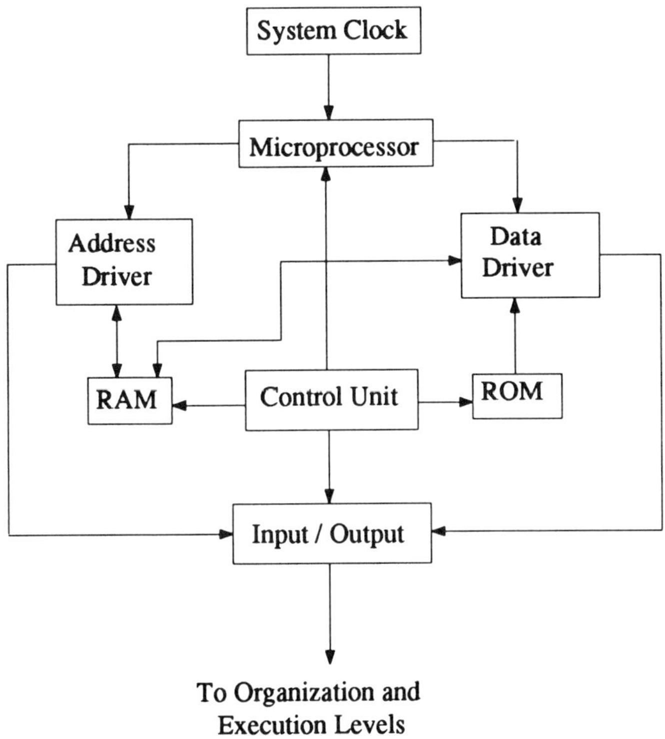

Figure 5.10 Hardware Coordinator Configuration

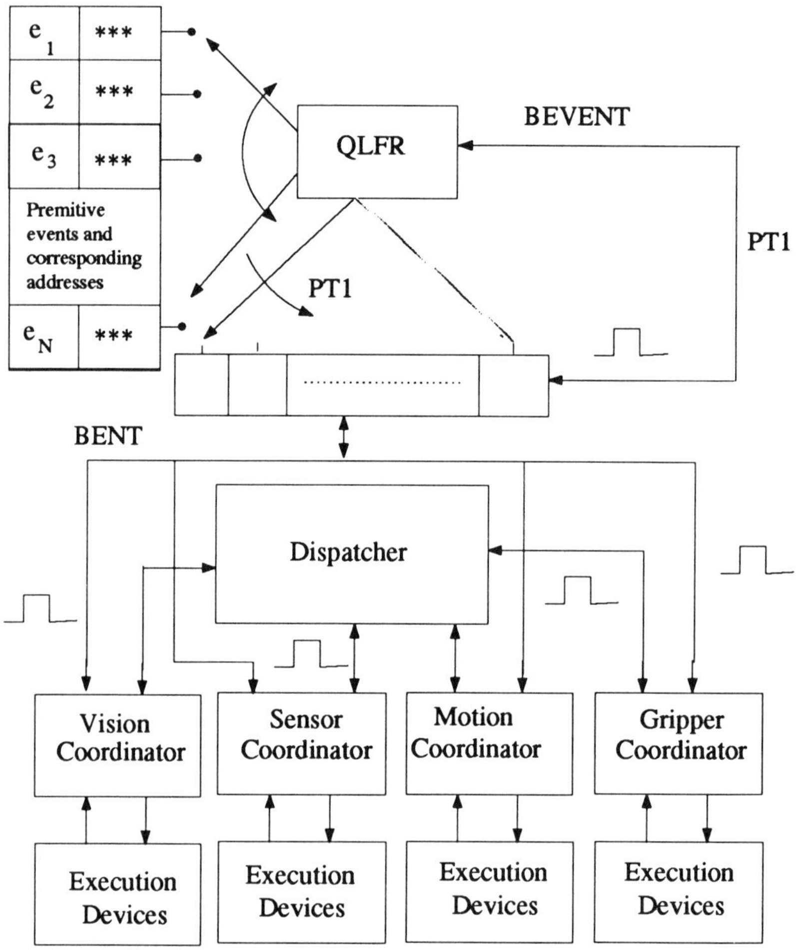

Figure 5.11 Schematic for the Coordination Level

When a specific coordinator completes its specific functions and the on line feedback information has (already) been communicated to it, this information is stored in the dispatcher memory of the coordination level. It may be used by other coordinators to complete their functions (during the execution of a requested job) and/or to calculate the overall accrued cost associated with the coordination level, that will be communicated to the organizer after the execution of the job. Thus, the different coordinators do not communicate directly with each other. But each coordinator has access to the dispatcher for storage and retrieval of data.

The coordination level does not have a particular memory equivalent to the steady state D_{ss} memory of the organization level. This is because the workspace environment of the coordination level is dynamic: Given a final plan Y^F, the formulation of the actual control problem regarding the way of its execution depends not only on previous experience but also on the current configuration of the workspace environment. For example, previously chosen trajectories for the different motions of the manipulator(s) should be modified by the presence of more or less obstacles and/or additional objects.

5.4 The Architectural Model for the Execution Level

The execution level executes the commands issued by the coordination level. Each control problem is analyzed in the light of its special requirements. Therefore, there is not one general architectural model that can include every possible operation performed at the execution level.

But although this is the case, the execution level consists of a number of devices each associated with a specific coordinator and vice versa. Each device is accessed via a command issued by its coordinator. Hence, the hierarchical structure is preserved.

In the case of an Intelligent Robotic System the execution devices at the execution level are the components of the vision system, the different sensors and the manipulators with their corresponding grippers. Consider for example the motion system coordinator. When a command is issued for a specific motion, the robot arm controller applies direct inputs to each of the joint actuators to move the arm to the desired final position.

5.5 The Architectural Model for the Upgrade Phase

The architectural model for the upgrade phase of Intelligent Robotic Systems includes specific hardware units that must be built within each of the higher two levels to account for both types of feedback information, on line and off line feedback.

The off line feedback mechanism (from the coordination to the organization level) is activated after the execution of the requested job and is used to upgrade the probability distribution functions of the sets of pertinent, ordered and augmented ordered activities associated with a compiled input command, as well as the knowledge base of the organizer.

The probabilities upgrade algorithm is given by the following equation as already explained:

$$p(t+1/u_j) = p(t/u_j) + \beta_{t+1}[\xi - p(t/u_j)] \qquad (5.5.1)$$

Each upgrade requires one addition, one subtraction and one multiplication. During this process there are no data dependencies between probabilities. According to [76, 77], a special purpose hardware unit is the one shown in Figure 5.12. It consists of two adders, one multiplicator unit and three registers. The probability value is fetched from the memory, passes through the three arithmetic operations and is returned to the memory, overwriting the previous value. This function is performed after the execution of the requested job and calculation of the overall accrued cost associated with the coordination level.

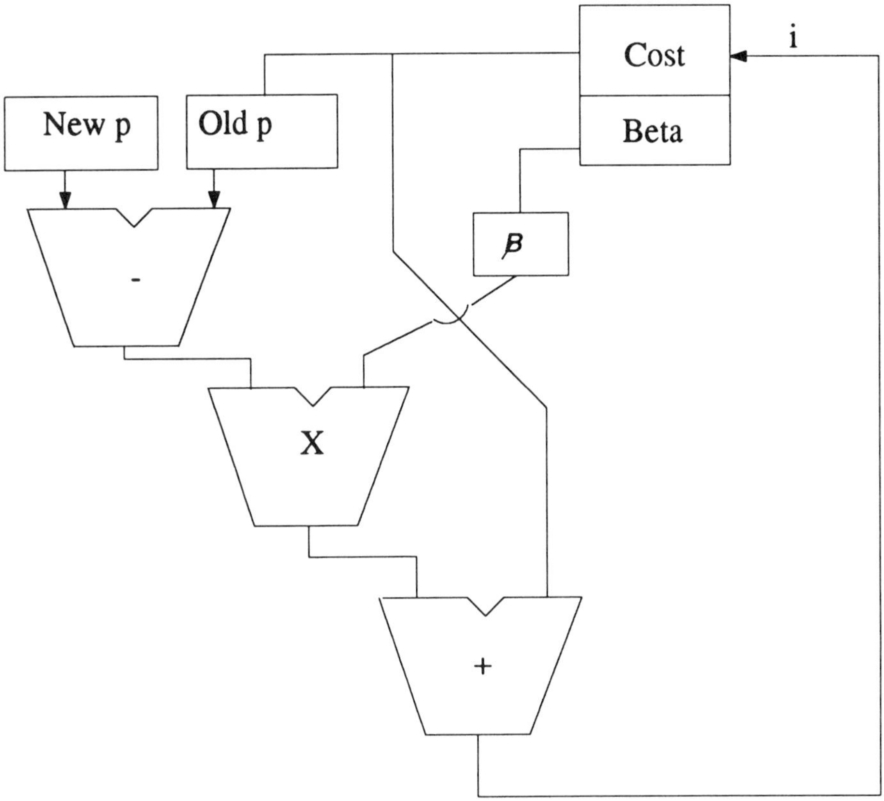

Figure 5.12 Probability Upgrade Hardware Unit

The on line feedback mechanism (from the execution to the coordination level) is activated during the execution of the requested job. A block diagram of this mechanism is shown in Figure 5.13. Solid lines represent the on line feedback information from the execution devices to the different coordinators. Individual and accrued costs are calculated within each coordinator and are transferred to the dispatcher of the coordination level where the overall accrued cost associated with the coordination level is calculated and communicated back to the organizer after the execution of the job. The dotted lines illustrate how the information from the dispatcher is used by the different coordinators and their execution devices for the completion of the job.

5.6 Remarks

The architectural model presented in this chapter has three major advantages:

1. It is applicable to any type of machine operating under the constraints of hierarchically intelligent control systems.
2. The same architecture may be used if one uses a linguistic approach to design such machines. Based on the architectural model described, one may derive the context free grammars that satisfy the same performance requirements and criteria, and,
3. Graph theory may be used to expand the architectural model by considering direct serial interaction between the coordinators.

Context free grammars are sufficient languages that fit perfectly well to the design of Intelligent Robotic Systems with hierarchical tree structures. Otherwise, more complicated languages like Webb grammars are necessary for similar designs with different structures. The architectural model of the organization level may be also used if one wishes to modify the machine reasoning function in the algorithm of the organization level.

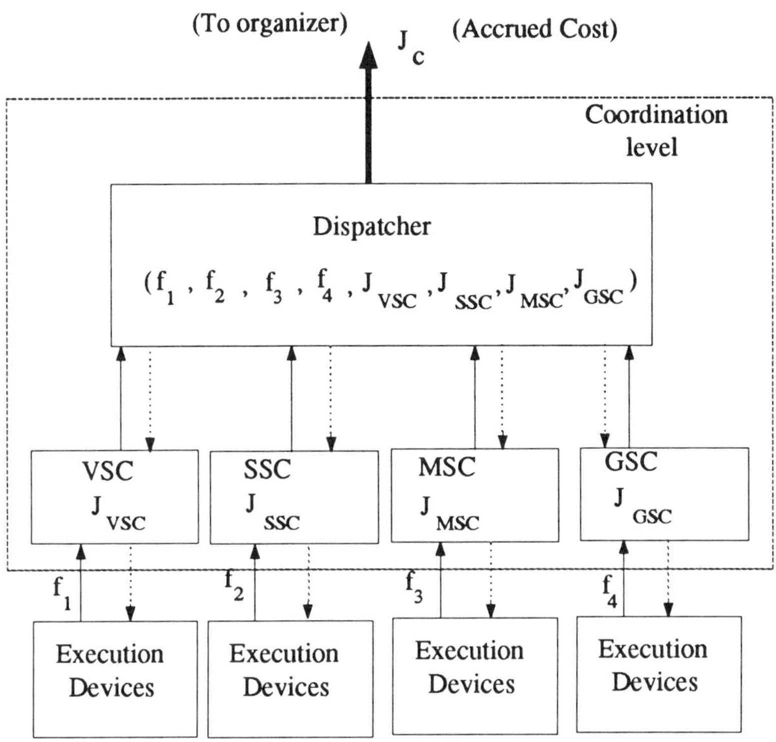

Figure 5.13 The Architectural Model for the On-line Feedback Mechanism

Given a compiled input command u_j, associate with it only a subset of the $(2^N - 1)$ pertinent activities which contains only those activities with corresponding probabilities greater than a prespecified $P_{\min}(u_j)$. Therefore, the machine planning and decision making functions are limited to the formulation of complete and compatible plans that originate from this subset of the $(2^N - 1)$ possible pertinent activities. This approach requires, decision making from the beginning: the machine reasoning function eliminates every activity with corresponding probability less than $P_{\min}(u_j)$.

The disadvantage of the architectural model is that it does not consider explicitly the possibility of unpredictable events during plan(s) execution. Further research in this area is needed and it is suggested as a topic of future research. Moreover, the integrated diagnostics and control architecture must be integrated with the rest of the system.

The next chapter presents some examples which justify the proposed models and architectures.

Chapter 6
EXAMPLES AND CASE STUDIES

6.1 Introduction

The purpose of this chapter is to illustrate and implement the probabilistic model and expert system cell derived for Intelligent Robotic Systems to some case studies of different complexity. Three case studies are studied:

1. An Intelligent (Mobile) Robot Operating System (IROS) designed to overcome emergency situations in a nuclear power plant.
2. A Robotic Assembly (RA) system designed to manufacture a mechanical part out of four smaller parts coming at random on a conveyor belt, and,
3. An Intelligent Robotic System operating in a hospital room environment.

The probabilistic approach is used for the first two examples. An expert system cell is used for the third example. Results have been obtained by simulations, considering realistic constraints. The model for integrated control and diagnostics, presented in chapter 4, is not considered in these examples. That is, it is assumed that all system components function properly. Different points are emphasized in each case study, so that all important aspects of the models and algorithms are illustrated, one way or another. All details associated with the case studies may be found in [237], [236] and [235], respectively.

6.2 An Intelligent Robot Operating System for Emergency Situations In A Nuclear Power Plant

As the first case study, we investigate a nuclear robotics problem related to emergency situations in a nuclear power plant environment with a Pressurized Water Reactor (PWR). The autonomous mobile robotic system is activated when emergency situations occur within the main reactor and it is expected to function in place of human operators and personnel.

Our preference to this particular problem has been based on the following three reasons:

1. A nuclear plant is a hazardous environment for humans because of its highly radioactive nature and may cause serious and sometimes fatal problems to their health. Operations during emergency situations (at least) should be performed by (smart) robots.
2. A very similar problem has been studied from a different point of view (Artificial Intelligence approach) by the Center for Engineering Systems Advanced Research (CESAR), established at the Oak Ridge National Laboratory (ORNL) by the U.S. Department of Energy [21, 22, 25, 26, 161, 216], and,
3. A number of different countries besides the United States (Japan, France, West Germany and Norway) have already developed and/or are developing various types of specialized robots for con-

ducting regular and repetitive work activities during periodic inspection within a nuclear power plant [173].

A complete and detailed methodology to overcome emergency situations by using autonomous robots is yet to be developed. Thus, by implementing our approach to this illustrative and realistic case study, we provide one way to solve this problem.

Currently, Japan, France and the United States are the only countries (world wide) that already use and/or are planning to use specialized robots in nuclear power plants. Therefore, an overview of what has been accomplished thus far, and what has been planned for the near future, will provide sufficient information about the capabilities of a (smart) robot, and, support further our belief for nuclear robotics.

Review on Nuclear Robotics Research

It has been mentioned that nuclear power plants are hazardous and frequently highly radioactive environments for humans. Every work task undertaken by humans in such environments involves exposure to radioactive materials which may cause fatal health problems.

In order to cope with such situations, a number of different countries like Japan, France, Germany, Norway and the United States have suggested the development of automation, remote control of maintenance work and development of smart robots capable of performing complex work tasks at various places of the nuclear facility. Japan and France already use robots for maintenance work within nuclear power plants while the United States is in the process of doing the same thing. The three countries are also developing intelligent manipulators for the future. A review of their research program is presented in the sequel.

Nuclear Robotics in Japan

The development of automation and remote control of maintenance work within nuclear power plants in Japan was advocated since the early 1970's in order to reduce the failure time of a reactor, uplift plant availability, improve the reliability of maintenance work and reduce the exposure of workers to radiation.

The present situation of operational robots and other autonomous equipment in both types of Japanese nuclear power plants, Pressurized Water Reactor (PWR) and Boiling Water Reactor (BWR), is summarized below [173]:

For PWR reactors:

1. Reactor vessel interior wall inspection equipment.
2. Steam generator robot.
3. Spent fuel inspection equipment.
4. Inspection equipment for the reactor vessel and other components.
5. Pipe cutting equipment.
6. Steam generator manipulator.
7. Steam generator in tank decontamination equipment, and,
8. Pressurizer heater replacement equipment.

For BWR reactors:

1. Main steam isolation valve automatic wrapping equipment.
2. High speed automatic fuel exchange equipment.
3. Control rod drive automatic exchange equipment.
4. Automatic cask cleaning equipment.
5. Self automatic cask cleaning equipment.
6. Inspection equipment for the reactor vessel and other components.
7. Automatic wall surface decontamination equipment.
8. Remote controlled and automatic in service inspection equipment, and,
9. Semi automatic disassembly inspection equipment.

The manipulators and all other autonomous equipment are currently completing tasks related to one of the following [173]:

1. Inspection, monitoring and repair works at areas of high level radioactivity such as those inside the containment vessel during plant operation.

2. Regular inspection, i.e., inspection of dimensional conformity, sound, temperature and radiation dose rate.
3. Transmission of precise data and image at abnormal occasions, i.e., observation of leakage, corrosion, damage, vapor, smoke, sound, vibration, and other conditions.
4. Simple work tasks like opening and closing of valves, tightening of gland, installing of measurement instruments, etc., and,
5. A number of similar tasks undertaken during periodic inspection, like disassemble and inspect valves, pumps and heat exchangers, change filters, decontaminate floors surrounding radioactive equipment, etc.

In addition to the above, the Japanese have recently developed a Remote Monitoring System for Inside Containment [173] composed of three subsystems, a floor-moving monitoring device, a rail-mounted monitoring device and a mobile manipulator. It allows early detection and response to emergency situations by the provision of detailed information concerning with the primary machines and tools installed in the container of nuclear reactors.

Further research related to the Japanese Project on Advanced Robot Technology to improve the plant availability and safety, includes [173]:

1. Locomotion on floors: Technology allowing four legged dynamic locomotion, incorporating a perceptive control system.
2. Locomotion on walls: Development of mechanisms consisting of two negative pressure adherences and wheeled locomotion structures joint with flexible articulations.
3. Data processing for control.
4. Multi finger manipulators.
5. Better actuators.
6. 3–D vision and visual information processing.
7. Optical transmission.
8. High reliability, and,
9. Radiation hardness.

The ultimate goal is to fully automate the Japanese nuclear power plants by the year 2000!

Nuclear Robotics in France

Research in nuclear robotics in France originated from efforts by E.D.F. (Electricite de France) to autonomize inspection and maintenance within nuclear power plants. This initiative was further supported because of the long term (and short term as well) advantages of fully automated nuclear environments [173]:

1. Reduction of the exposure time to radioactive materials of personnel.
2. Reduction of the time required to perform various tasks, and,
3. Increase of reliability and availability.

Currently, a robot named "ISIS" manufactured by H.P. and designed for E.D.F., is used to repair the French nuclear reactor (plant) at CHINON 3 (cooled gas). The ISIS robot is a heavy duty 8 axis robotic or remote manipulator used for inspection, repair, decontamination, decommissioning and dismantling in a nuclear environment. Depending on the type of work to be performed, this robot is capable of operating under three different modes:

1. Remote controlled manipulator mode, using a master controller.
2. Automatic mode commanded by a computerized prerecorded path, and,
3. Automatic mode using preprogrammed routes and final positioning using proximity and force feedback sensors.

The ISIS robot has the capability to interchange toolheads. This allows a variety of tasks to be performed and carried out in real time. These tasks include:

1. Survey of areas to be repaired, using a video camera range finder (telemetric head).
2. Display of areas to be repaired on a full scale mockup using a laser beam duplicating head.

3. Scouring of parts to be welded together with final inspection of surface finish.
4. Handling and positioning of parts, and,
5. Electric welding.

The above maintenance work can be detailed on the full scale mock up and the steps carried out by the operator are automatically programmed into the computer's memory.

In addition to the ISIS robot a "PIADE" machine is on use for hot cell dismantling (since April, 1982). PIADE is a moveable tool carrier for remote video controlled cutting and handling operations, adapted to hot cell dismantling.

Further research related to the Project De Robot D' Intervention Nucleaire De Cea includes the development and navigation of mobile robots, coordination of multi robots working simultaneously, anti collision techniques for vehicles, interface of robot controllers to expert systems and development of complex robotic systems architectures.

Nuclear Robotics in the United States

Research on nuclear robotics in the United States is performed at the Center for Engineering Systems Advanced Research (CESAR), established at the ORNL by the Department of Energy. The Center has addressed its long range, energy related research in intelligent control systems, intended to plan and perform a variety of tasks in unstructured hostile environments, given only qualitatively specified goals. Research objectives stress the minimization of risks associated with the production and utilization of (nuclear) energy. Potential applications will be in emergency situations and remote operations in hostile environments with major emphasis in nuclear power plants, resource exploration, and many others. In particular, activities have been initiated in the areas of mathematical modeling of robot dynamics, realtime control, world models, machine perception and strategy planning [21, 22, 25, 26, 161, 216].

As a first sample problem, they are developing smart mobile robots to operate in nuclear power plants when emergency situations occur and the normal electronic remote control of plant valves is impossible. This sample problem has been chosen as our first case study. A description of the problem along with the strategy formulation for its solution are presented in the sequel. Simplifications (when possible) are considered and justified to reduce the complexity of the problem. A step-by-step analysis of the problem is presented and two representative examples are given. All details may be found in [237].

Problem Statement

Assume that a nuclear power plant with a PWR shown in Figure 6.1 (modified in Figure 6.2) encounters an emergency situation. A power failure makes the normal electronic remote control of plant valves impossible. Subsequently, the pressure control in this highly radioactive environment fails, so that the negative pressure in the hazardous cell is no longer maintained. A breach in the primary containment (often several feet thick wall of concrete and steel, with penetrations for piping and manipulators from the working area occupied by operating personnel) is identified and operating personnel are evacuated.

Several valve-related, flange-related and pipe-related operations are required to be performed. An autonomous mobile robot named "Ralph" with visual capabilities is used to perform these operations.

There are three different types of valves within the PWR: relief, safety and isolation valves. *Relief valves* regulate the flow of steam from the Pressurizer and Steam Generator through the reactor. *Safety valves* regulate the rate of flow of steam. *Isolation valves* isolate certain pipes when they need repair and/or replacement, and during emergency situations (when necessary). Flanges are used to hold pipes in place and connect adjacent pipes.

When an emergency occurs, related operations that must be performed by Ralph are (in ascending order of complexity):

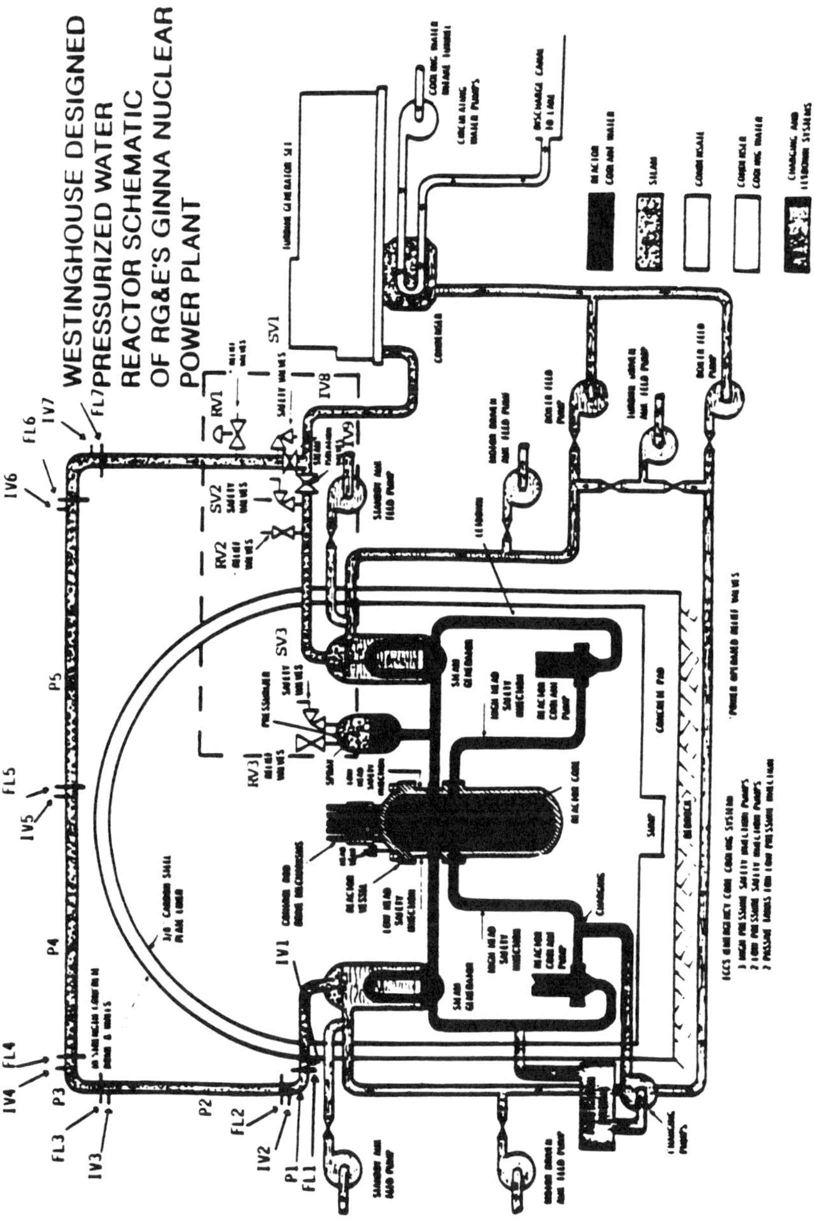

Figure 6.1 The PWR Nuclear Power Plant

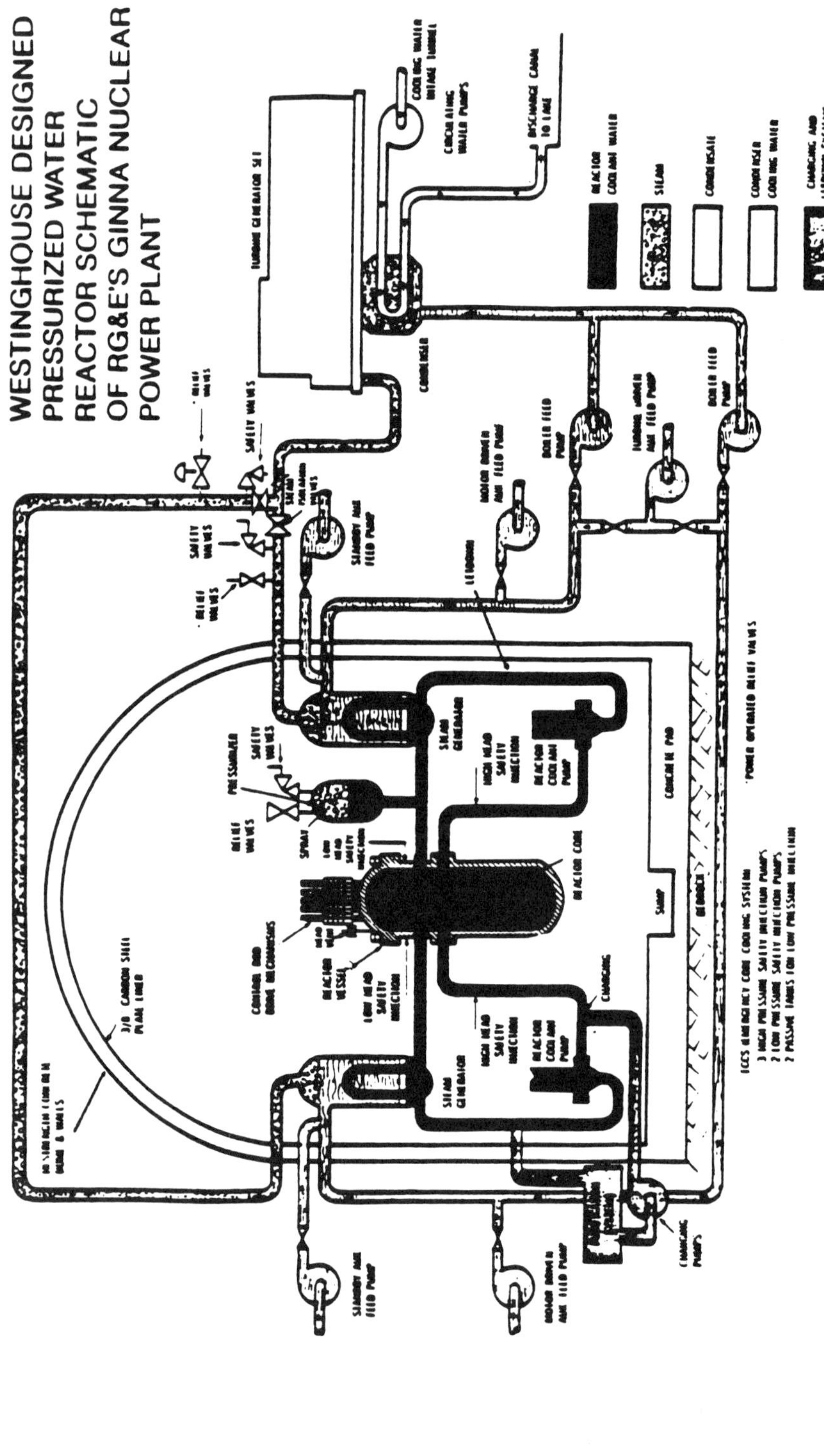

Figure 6.2 The Modified PWR Nuclear Power Plant

Valve related operations. They include: activation (on/off), repair and/or replacement of faulty valves and inspection to determine the current operability of the valves.

Flange related operations. They include: repair and/or replacement of faulty flanges (flanges which cause leakages) and inspection to determine the current operability of the flanges.

Pipe related operations. They include: inspection of pipes, repair and/or replacement of faulty pipes (pipes which cause leakages).

The overall problem is highly stochastic in nature due to the following facts:

1. The optical properties of the valves, flanges and pipes are unknown at the time of the emergency situation; i.e., valves, flanges and pipes may be clean or greasy.
2. The valve(s) (flange(s), pipe(s)) Ralph will operate on, is one of multiple valves (flanges, pipes) of this type located in this vicinity.
3. The operability of the valves, flanges and pipes at the time of the emergency is unknown (they may be corroded, slippery, etc.), and,
4. There are several paths which the mobile robot may follow through rooms, corridors and aisles, some of which may be obstructed by unanticipated objects at the time of the emergency.

Although the overall problem is highly stochastic, several justifiable simplifications may be considered to simplify its complexity for strategy formulation during an emergency situation.

A pipe-related operation requires both valve– and flange– related operations. For example if the user command is REPLACE PIPE #1, it is necessary to isolate PIPE #1 by closing the nearest isolation valves (activate them to interrupt the steam flow), remove the flanges which hold the pipe, remove and replace the pipe, tie the flanges and deactivate again the isolation valves. A flange-related operation requires valve-related operations but not pipe-related operations. Consider for example, a user command REPLACE FLANGE #2. It is necessary to isolate the flange, first, by activating the isolation valves (interrupt the steam flow) and then replace it. A valve-related operation is the simplest compared to the others and does not require either flange or pipe-related operations.

Therefore, when an emergency situation occurs and a command is classified either as a valve-related or flange-related or pipe-related operation, appropriate actions are taken to complete the requested job. The autonomous mobile robotic system is expected to receive the user command, interpret it and formulate suitable scenarios to complete the requested job. It is expected to learn from "experience" (as Ralph continues to function) and adapt to different situations.

The assumptions made for this sample problem are stated and clarified in the sequel. We explain the overall strategy formulation when an emergency occurs, and define the model of the workspace environment Ralph will operate in.

Figure 6.1 shows the Westinghouse designed PWR schematic of the RG&E GINNA nuclear power plant. The figure shows the location of the safety, relief and isolation valves of the Pressurizer and Steam Generator. But it does not show the actual pipe connections and location of flanges. Therefore, in order to make the case study much more realistic we consider a number of flanges at different locations within the nuclear power plant used for piping connections between the two Steam Generators and we introduce additional isolation valves associated with the flanges. These additions are shown in Figure 6.2. The dotted area is the steam regulation area where all valves regulate the steam flow from the Steam Generator. Based on the above additions and the modified nuclear power plant schematic shown in Figure 6.2, the following assumptions are considered in this case study:

1. The mobility problem is assumed to be solved; the corridors, rooms and aisles have been designed for mobile robots like Ralph.
2. Detailed built drawings of the PWR are not available. Therefore, the actual location of the valves, flanges and end points of the pipes is not known. What is known is that all valves flanges and

pipes are numbered clockwise, and the position of the first and last flange and valve.

3. It is assumed that Ralph *knows* that relief and safety valves are located close to the Pressurizer and Steam Generator (see the dotted area in Figure 6.2) and that isolation valves are mainly associated with piping connections.
4. There is a complete description of the valve types, flanges and pipes. Images of the valves, flanges and pipes may be generated from any angle.
5. The optical properties of the valves, flanges and pipes are unknown at the time of the emergency, i.e., they may be clean or greasy.
6. Access to the valves, flanges and pipes may be partially obstructed by unanticipated objects. Ralph must choose from several paths through corridors and aisles.
7. Time is very important for successful task completion.
8. The initial position of the robot arm is known. After the completion of a job, Ralph returns to its initial position, and,
9. Ralph's mechanical arms have six degrees of freedom each for manipulation, which are sufficient to allow operations on objects with any orientation in the three dimensional space. Ralph may also elevate at all desirable levels within the plant, to reach valves, flanges and pipes.

The second assumption is very important. It complicates the problem because it considers that Ralph is responsible to identify and operate on the proper valves flanges and pipes. On the other hand, it provides Ralph with some *hints* (partial knowledge (information)) about the possible location of a valve, flange or pipe, since the location of the first and last valve and flange is known, and pipes are located between flanges. The third assumption provides Ralph with some additional information: When an operation is requested on a safety or relief valve, the search area is not the whole nuclear power plant, but the dotted area near the Pressurizer and Steam Generator. But even in this simpler case, there is an uncertainty associated with the strategy formulation by Ralph, because the exact location and operability of the specific valve(s) is unknown. Another very important assumption that has not been stated explicitly is that Ralph is expected to learn from experience as he continues to function within the nuclear power plant. This means that when Ralph faces the same problem more than once, he is expected to be more *knowledgeable* after the first time. This is accomplished by storing and upgrading all information in the pertinent memory. Consider for example the very simple case where Ralph is requested to operate on the same valve (flange or pipe) for a second time. The location of the valve (flange or pipe) under consideration is already known to Ralph from the first time. Therefore, there is no need to identify and recognize the proper valve (flange or pipe). What is still needed, is information about the operability and how he will have access to the valve (flange or pipe). With this procedure, Ralph will reach a critical point where the location of every valve and flange and the end points of the pipes will be known to him. Thus, Ralph will be eventually aware of the detailed built drawings of the PWR plant. However, the uncertainties related to its strategy formulation may still exist; due to the emergency situations, the model of the workspace environment is always different, i.e., previously available paths are cut off, access to the valves, flanges and pipes may be obstructed by more or less objects, etc.

In addition to the above assumptions, it is necessary to clarify two more aspects: The first one is related to operations on a valve and/or flange. Ralph is assumed to know a priori (based on the received command) if he must rotate the valves and/or flanges he operates on clockwise or counterclockwise. This is the only realistic approach to the problem since any experimental technique may cause a disaster within the nuclear plant! The second aspect has to do with when an emergency situation is declared within the plant. An emergency situation is declared within a nuclear power plant when a flaw is sensed. Radioactive visual and non visual sensors are considered at different locations within the plant. Their purpose is to measure continuously the radiation and contamination levels, and make sure that they are within the acceptable limits for the operating personnel. If and when the radioactive sensors sense a (minor or major) flaw, the operating personnel are evacuated, Ralph is activated and is waiting for a command in order to function.

Strategy Formulation

There are three different types of user commands that may be issued to Ralph during an emergency situation. These commands are related to valves, flanges and pipes operations. Given a user command, Ralph formulates different strategies to execute the requested job. The overall strategy formulation for each type of user commands (valve, flange or pipe related commands) is more or less the same, regardless of the specific requested operations on a valve, flange or pipe. The overall strategy formulation procedures associated with each type of user commands are described below.

Given a valve-related command, Ralph is expected to formulate a strategy to:

1. Locate and identify the proper valve (or valves if operations are requested on more than one).
2. Approach the valve(s) considering possible paths and obstacles that may partially obstruct Ralph's access to the valve(s).
3. Identify the valve's origin with respect to its mobile platform and elevate at the appropriate level (to operate on the valve).
4. Evaluate the operability of the valve by inspection.
5. Find the position and orientation of the handle (strap) part in order to grasp it, and,
6. Complete the job and return to the initial position.

For a flange-related command, the strategy formulation is more complicated. After the location and identification of the proper flange(s) (similar action to the one taken for a valve operation at step 1, it is necessary to locate, identify and activate the isolation valve(s) associated with the specific flange(s). Once this is done (the steam flow is interrupted), a strategy similar to the one for a valve operation is formulated (steps 2-5). The last step is extended to include deactivation of the isolation valve(s) before Ralph returns to its initial position.

The strategy formulation for a pipe-related command is the most complicated of all three. After the identification of the proper pipe and location of its end points, the flanges and isolation valves associated with the specific pipe are also identified. The isolation valves are activated to interrupt the steam flow and operations of the flanges are performed before any operations to be performed on the pipe. A similar strategy procedure is also followed in this case (steps 2–5). The last step is further extended to include deactivation of the isolation valves and flanges before Ralph returns to its initial position.

Organization and formulation of possible complete and compatible plans is done at the organization level, as previously described. The details are added after the selection of the most probable complete and compatible plan to execute the user requested job.

The next section describes the probabilistic model of the workspace environment Ralph will operate in. A number of realistic simplifications are considered which reduce the complexity of the model. These simplifications are introduced and justified one-by-one. The final model is shown to be sufficient for the case study.

The Model of the Workspace Environment and Related Simplifications

The model of the workspace environment is derived based on the modified schematic of the pressurized water nuclear power plant shown in Figure 6.2. We consider emergency situations that may occur within the main reactor due to malfunctions of the Steam Generators and Pressurizer, and emergency situations that may occur due to leakage in the pipes which connect the two Steam Generators. Based on the figure we consider:

1. Three safety valves *SV1* to *SV3* (within the dotted area) near the Pressurizer and Steam Generator, used to regulate the flow of steam.
2. Three relief valves *RV1* to *RV3* (within the dotted area) near the Pressurizer and Steam Generator, used to regulate the rate of flow of steam.
3. Nine isolation valves *IV1* to *IV9*. Seven of them (*IV1* to *IV7*) are used to interrupt the steam flow through the pipes which connect the Steam Generators. The last two (*IV8* and *IV9*) are used for emergency situations caused by a flaw in the Pressurizer and/or right Steam Generator.
4. Seven flanges *FL1* to *FL7* used to connect adjacent pipes. They are located at the end points of the different pipes, and,
5. Six pipes *P1* to *P6* used to connect the two Steam Generators.

Every valve (isolation, relief, safety), flange and pipe is numbered clockwise. The isolation valves *IV1* to *IV7* outside the dotted area are associated with the flanges *FL1* to *FL7*. One and only one isolation valve is associated with each flange: *{ IVi <–> FLi, i=1,2,...,7 }*.

Since the flanges (and therefore their corresponding isolation valves) are located at the end points of the pipes, there are two flanges (and two isolation valves) associated with each pipe: *{ Pi <-> ((FLi, FL(i+1)), (IVi, IV(i+1))), i=1,2,...,6 }*.

Therefore, for flange-related (pipe-related) operations, once the proper flange (pipe) has been located and identified, the search area for the location and identification of the corresponding valve (flanges and valves) is restricted. For operations on the relief, safety and the last two isolation valves, the search area is restricted to the dotted area near the Pressurizer and Steam Generator.

The three types of user commands that are considered during emergency situations are:

TYPE I: Commands for valve-related operations:

c_1 = Regulate-more *RV1*

c_2 = Regulate-more *RV2*

c_3 = Regulate-more *RV3*

c_4 = Regulate-more *SV1*

c_5 = Regulate-more *SV2*

c_6 = Regulate-more *SV3*

c_7 = Deactivate *IV8*

c_8 = Deactivate *IV9*

c_9 = Regulate-less *RV1*

c_{10} = Regulate-less *RV2*

c_{11} = Regulate-less *RV3*

c_{12} = Regulate-less *SV1*

c_{13} = Regulate-less *SV2*

c_{14} = Regulate-less *SV3*

c_{15} = Activate *IV8*

c_{16} = Activate *IV9*

TYPE II: Commands for flange-related operations:

c_{17} = Replace *FL1*

c_{18} = Replace *FL2*

c_{19} = Replace *FL3*

c_{20} = Replace *FL4*

c_{21} = Replace *FL5*

c_{22} = Replace *FL6*

c_{23} = Replace *FL7*

TYPE III: Commands for pipe-related operations:

c_{24}	=	Replace *P1*
c_{25}	=	Replace *P2*
c_{26}	=	Replace *P3*
c_{27}	=	Replace *P4*
c_{28}	=	Replace *P5*
c_{29}	=	Replace *P6*

"Regulate-more" implies a valve operation that results in an increase of the flow, or rate of flow of the steam through the pipes - counterclockwise rotation of the valves. "Regulate-less" means the opposite; clockwise rotation of the valves.

"Activate" implies interrupt the steam flow by closing completely the isolation valves. "Deactivate" means the opposite.

Thus, the set C, of user commands contains M=29 different commands; i.e., $C = \{c_1, c_2, ..., c_{29}\}$. The number of the (classified) compiled input commands to the organization level is also M=29; i.e., $U = \{u_1, u_2, ..., u_{29}\}$. In this case study we consider an *ideal* classifier so that there is a one-to one correspondence between the user and compiled commands. That is, $c_i <-> u_i$, $p(c_i) = p(u_i)$.

The task domain contains N=18 (repetitive and non repetitive) primitive events; i.e., $E_t = \{e_1, e_2, ..., e_{18}\}$. Two of them ($L$=2) are repetitive primitives, $E_r = \{e_{17}, e_{18}\}$, while sixteen of them (N-L=16) are non repetitive ones, $E_{nr} = \{e_1, e_2, ..., e_{16}\}$. The primitive events are:

e_1	=	Return (to initial position)
e_2	=	Grasp valve handle and rotate clockwise
e_3	=	Grasp valve handle and rotate counterclockwise

e_4	=	Grasp valve handle and rotate clockwise *IV* valve
e_5	=	Grasp valve handle and rotate counterclockwise *IV* valve
e_6	=	Grasp valve handle and rotate clockwise *IV* valves
e_7	=	Grasp valve handle and rotate counterclockwise *IV* valves
e_8	=	Grasp flange strap and tie
e_9	=	Grasp flange strap and untie
e_{10}	=	Grasp flanges strap and tie
e_{11}	=	Grasp flanges strap and untie
e_{12}	=	Grasp flange
e_{13}	=	Dump flange
e_{14}	=	Grasp pipe
e_{15}	=	Dump pipe
e_{16}	=	Replace
e_{17}	=	Move
e_{18}	=	Camera

The primitive events e_2, e_3 and e_4, e_5 denote operations on one valve, namely the relief or safety and the isolation valves *IV8* and *IV9*, respectively. The primitive events e_8, e_9 are used only for flange-related operations and indicate operations on one flange. On the other hand, the primitive events e_6, e_7, e_{10}, e_{11}, denote operations on two valves and two flanges, those located at the end points of a pipe, occurring when the requested jobs are associated with pipe-related operations.

The organization and formulation of plan scenarios without any of the proposed rules is extremely complex (as has been shown in [237]). Therefore, simplifications are necessary.

In order to justify further these simplifications and specify exactly the active and/or inactive primitive events associated with each user (or compiled) command, consider the set C (U) of the user (compiled) commands and the set E_t of primitive events. Observe that:

1. If the user (or compiled) commands are from c_1 to c_6, then potential activity candidates to formulate complete and compatible plans should contain only primitive events from the subset $E_t^1 = \{e_1, e_2, e_{17}, e_{18}\}$.
2. If the user (or compiled) commands are from c_9 to c_{14}, then potential activity candidates to formulate complete and compatible plans should contain only primitive events from the subset $E_t^2 = \{e_1, e_3, e_{17}, e_{18}\}$.
3. If the user (or compiled) commands are c_7 and c_8, then potential activity candidates to formulate complete and compatible plans should contain only primitive events from the subset $E_t^3 = \{e_1, e_5, e_{17}, e_{18}\}$.
4. If the user (or compiled) commands are c_{15} and c_{16}, then potential activity candidates to formulate complete and compatible plans should contain only primitive events from the subset $E_t^4 = \{e_1, e_4, e_{17}, e_{18}\}$.
5. If the user (or compiled) commands are from c_{17} to c_{23}, then potential activity candidates to formulate complete and compatible plans should contain only primitive events from the subset $E_t^5 = \{e_1, e_4, e_5, e_8, e_9, e_{12}, e_{13}, e_{16}, e_{17}, e_{18}\}$.
6. If the user (or compiled) commands are from c_{24} to c_{29}, then potential activity candidates to formulate complete and compatible plans should contain only primitive events from the subset $E_t^6 = \{e_1, e_6, e_7, e_{10}, e_{11}, e_{14}, e_{15}, e_{16}, e_{17}, e_{18}\}$.

These subsets are not necessarily disjoint but they must satisfy: $E_t^1 \, U \, E_t^2 \, U E_t^3 U E_t^4 U E_t^5 U E_t^6 = E_t$

Therefore, for valve-related operations, a maximum number of two active non repetitive primitives is sufficient for plans organization and formulation, while for flange and/or pipe-related operations a maximum number of eight active non repetitive primitives is sufficient. But although the number of primitives for flange and/or pipe-related operations is the same, pipe-related operations require operations on two valves and flanges. Thus, the order of complexity during plans formulation, is preserved.

It is also necessary to clarify that even though the same subset of primitive events E_t^i, $i = 1, 2, ..., 6$, may be used to formulate complete plans for a group of commands (i.e., E_t^1 for c_1 to c_6), the cost of execution of the most probable plan is different for each command, because each command requires operations on different components (valves, flanges, pipes) within the nuclear power plant.

Based on the above problem reduction and justified simplifications, given a user or compiled command (ideal classifier), the probabilities of the primitive events which belong to the sets $E - E_t^i, i = 1, 2, ..., 6$, are set by default to zero, since the corresponding primitive events e_i are irrelevant for plans organization. Every other primitive event, may be either active or inactive with a certain probability within the respective activity candidates. The number of activities is reduced further when one considers the compatibility tests. Furthermore, it is straightforward to define all sets of events shown in Table 1, chapter 4.

To be more specific, in all cases, $E_{end} = \{e_1\}$, $E_{sr} = \{e_{18}\}$ (assuming that Ralph always needs a visual scanning of the workspace environment), $E_{ord} = \{(e_{18}\, e_{17})\}$, the only valid sequence of repetitive events; there is a visual scanning at first and then a potential motion, while $E_c^i = E_t^i - E_r$, $i = 1, 2, ..., 6$. The pertinent set of critical non repetitive events E_{cr}^i are obtained by taking all valid permutations of E_c^i, Per$\{E_c^i\}$, checked against specific pairs of compatible events. Finally, every pair of the form $(e_1 e_i)$ i= 2, 3,...,18, belongs to the list of unwanted pairs of primitive events.

The pairs of compatible primitive events are given below, followed by the models for the coordination and execution levels.

$$E_{comp} = \{(e_{18}\, e_{17}), (e_2\, e_1), (e_3\, e_1), (e_4\, e_1), (e_5\, e_1), (e_4\, e_9)$$
$$(e_6\, e_{11}), (e_7\, e_1), (e_8\, e_5), (e_9\, e_{12}),$$
$$(e_{10}\, e_7), (e_{11}\, e_{14}), (e_{12}\, e_{13}), (e_{13}\, e_{16}),$$
$$(e_{14}\, e_{15}), (e_{15}\, e_{16}), (e_{16}\, e_8), (e_{16}\, e_{10}),$$
$$(e_{17}\, e_1), (e_{17}\, e_2), (e_{17}\, e_3), (e_{17}\, e_4), (e_{17}\, e_5),$$
$$(e_{17}\, e_6), (e_{17}\, e_7), (e_{17}\, e_8), (e_{17}\, e_9), (e_{17}\, e_{10}),$$
$$(e_{17}\, e_{11}), (e_{17}\, e_{12}), (e_{17}\, e_{13}), (e_{17}\, e_{14}), (e_{17}\, e_{15}),$$
$$(e_{17}\, e_{16}), (e_2\, e_{18}), (e_3\, e_{18}), (e_4\, e_{18}),$$
$$(e_5\, e_{18}), (e_6\, e_{18}), (e_7\, e_{18}), (e_8\, e_{18}), (e_9\, e_{18}),$$
$$(e_{10}\, e_{18}), (e_{11}\, e_{18}), (e_{12}\, e_{18}), (e_{13}\, e_{18}),$$
$$(e_{14}\, e_{18}), (e_{15}\, e_{18}), (e_{16}\, e_{18})\}$$

The Coordination and Execution Level Model

This section describes the workspace models of the coordination and execution levels and derives explicitly the individual and accrued costs associated with both levels. In the absence of any other information, a number of realistic assumptions has been, again, considered. They are presented and justified in the sequel.

The valves are assumed to have *spherical shape.* The flanges spherical-cylindrical and the pipes cylindrical. The individual cost functions associated with the recognition procedures are calculated based on the analysis presented in [179]. Their values are:

For valves: $J_1 = (1/k) * 20\,\text{nats}$

For flanges: $J_2 = (1/k) * 40\,\text{nats}$

For pipes: $J_3 = (1/k) * 30\,\text{nats}$

where k denotes the iteration number, i.e., how many times the same command has been issued. These costs are reduced as Ralph continues to function.

A reasonable *measurement resolution* of 0.005 degrees with a maximum *uncertainty interval* of 1.5 degrees is considered for the vision system. Therefore, the determination of a specific position has an associated uncertainty of 5.70 nats. Since Ralph is expected to be more knowledgeable as he continues to function, this uncertainty is reduced proportionally to the inverse of the iteration number *k*. The explicit formula for the determination of the position of a valve, flange and pipe is:

$$\text{For valves: } J_{1v} = (1/k) * 5.70 \text{ nats}$$

$$\text{For flanges: } J_{2f} = (1/k) * 5.70 \text{ nats}$$

$$\text{For pipes: } J_{3p} = (1/k) * 11.40 \text{ nats}$$

Therefore:

$$J_{1v} + J_{1v} = (1/k) * 25.70 \text{ nats}$$

$$J_2 + J_{2f} = (1/k) * 45.70 \text{ nats}$$

$$J_3 + J_{3p} = (1/k) * 41.40 \text{ nats}$$

J_{3p} is twice as large as the other two, because identification of two end points is necessary for pipe-related operations.

There are two paths Ralph may follow within the plant. The selection of the path depends on the feedback information about the configuration of the environment. Each path is chosen with a certain probability $p(W_{\Gamma\lambda})$, $\lambda = 1, 2$. The cost of each motion within the plant depends on the type of operation, i.e., valve, flange or pipe. It is considered fixed for each motion. Without loss of generality, assume that the corresponding values of the associated costs are *A1*, *A2* and *A3* $(A1 < A2 < A3)$, respectively. Each motion that corresponds to a specific type of operation, is penalized differently by a cost of *B1*, *B2*, *B3*, $(B1 < B2 < B3)$, respectively. Thus, three penalty functions are considered. A penalty function is chosen based on the specific operation, i.e., valve operation, with a certain probability $p(W_{Pd})$, $d = 1, 2, 3$. Based on the previous assumptions, the accrued cost of the Motion System Coordinator is:

For valve operations:

$$J_{MC} = \mu * A1 + B1 + H(p(W_{\Gamma\lambda})) + H(p(W_{Pd})) \qquad (6.2.5.1)$$

For flange operations:

$$J_{MC} = \mu * A2 + B2 + H(p(W_{\Gamma\lambda})) + H(p(W_{Pd})) \qquad (6.2.5.2)$$

For pipe operations:

$$J_{MC} = \mu * A3 + B3 + H(p(W_{\Gamma\lambda})) + H(p(W_{Pd})) \qquad (6.2.5.3)$$

where, $\lambda = 1, 2$, $d = 1, 2, 3$ and μ the number of movements during the execution of the plan. All other costs are considered negligible. The overall accrued cost of the coordination level is:

For valve operations:
$J_c = \mu * A1 + B1 + H(p(W_{\Gamma\lambda})) + H(p(W_{Pd})) + (1/k) * 25.70$ nats

For flange operations:
$J_c = \mu * A2 + B2 + H(p(W_{\Gamma\lambda})) + H(p(W_{Pd})) + (1/k) * 71.40$ nats

For pipe operations:
$J_c = \mu * A3 + B3 + H(p(W_{\Gamma\lambda})) + H(p(W_{Pd})) + (1/k) * 214.20$ nats

Based on the configuration of the plant, two possible kinds of feedback information are considered: f1 and f2. When f1 is received, path $W_{\Gamma 1}$ must be chosen and when f2, then $W_{\Gamma 2}$.

Initial Conditions

All initial probability distribution functions are assumed to be uniform. That is, the initial probabilities of the input commands are $p(u_j) = 1/29$, $j = 1, 2, ..., 29$. The initial entropy associated with each command is 0.116 infor. nats while the entropy rate is 0.116 information nats/step. The total initial entropy rate associated with all input commands is 29*H$(u_j) = 3.3673$ information nats/step.

In a similar way, all activities are assumed to have uniform initial distributions [237].

For the coordination level, $p(W_{\Gamma\lambda}) = 0.5$, $\lambda = 1, 2$ $p(W_{Pd}) = 0.333$, $d = 1, 2, 3$.

Two illustrative examples are considered next.

Examples

Two examples, one for valve operations and the other for flange operations are considered. All details for the plan formulation, organization and execution are presented in [237]. Therefore, a summary only is given in this section. The results are presented via a set of four Figures. The Figures illustrate respectively:

1. The evolution of the probability distribution function $P(X_{jm}/u_j)$ of the best initial activity (string) used to formulate the complete plan, the evolution of the probability distribution function $P(Y_{jmr}/u_j)$ of the best ordered activity (string) and the evolution of the probability distribution of the best plan candidate $P(B_{jmv}/u_j)$.
2. The evolution of the entropy rates of the previous strings.
3. The evolution of the entropy rates of the activated memory blocks D_j, D_{jj}, D_{jjj}, and,
4. The evolution of the total Entropy rate of the three activated blocks.

Example 1

Consider that the command is $c_4 = u_4 =$ "Regulate-more $SV1$". Once it is received and acknowledged by the organization level, the switch s_α is realized to activate the corresponding memory block D_4 and transfer its data to RB and to PRB. The block D_4 contains all relevant strings X_{4m}. Its entropy rate $\bar{H}(D_4)$ indicates the throughput rate from reasoning to planning. The rest 28 blocks denote the blockage rate, information blocked within machine reasoning. When D_4 is received by machine planning, the switch s_β is realized and the block D_{44} is activated to formulate compatible ordered strings. Realization of s_γ permits activation of the block D_{444} and formulation of augmented activities-strings by inserting valid sequences of repetitive primitive events and checking for computibility and completeness. Only one passes the completeness test. This string is sent to the coordination level, since decision making is not necessary in this case.

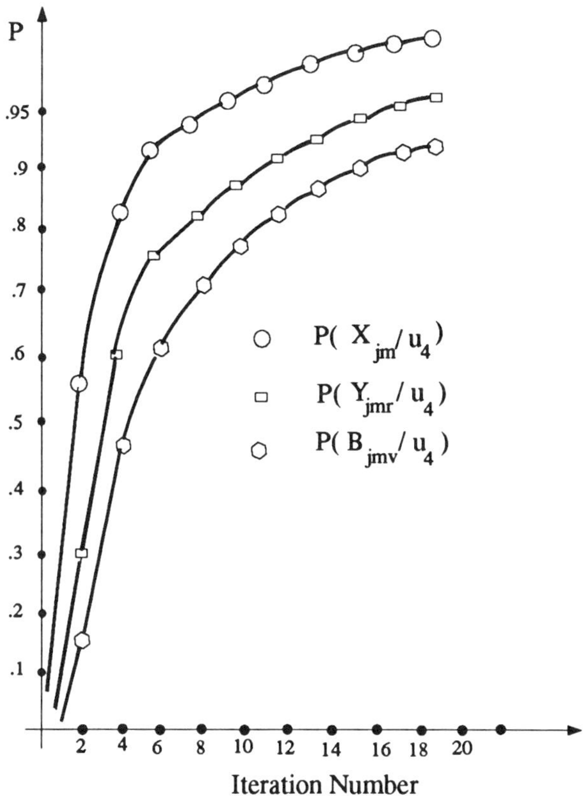

Figure 6.3 Evolution of Probability Distribution Function for Sequential Testing of $c_4 = u_4$

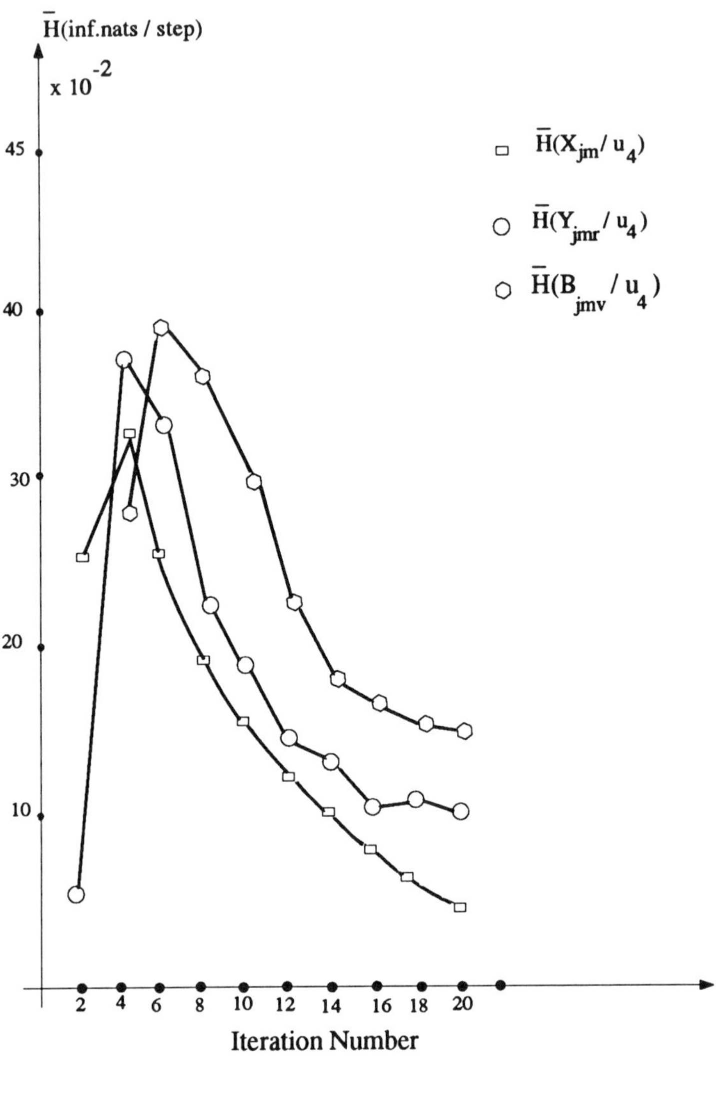

Figure 6.4 Evolution of Entropy Rates for Sequential Testing of $c_4 = u_4$

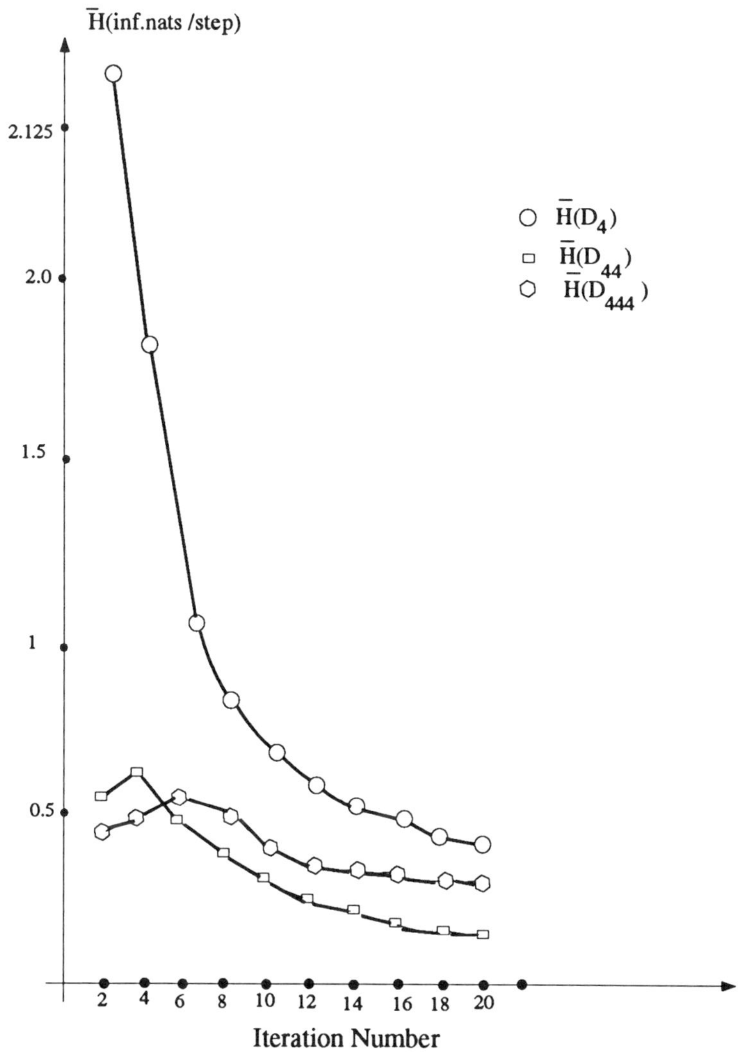

Figure 6.5 Evolution Rates of Activated Memory Blocks for Sequential Testing of $c_4 = u_4$

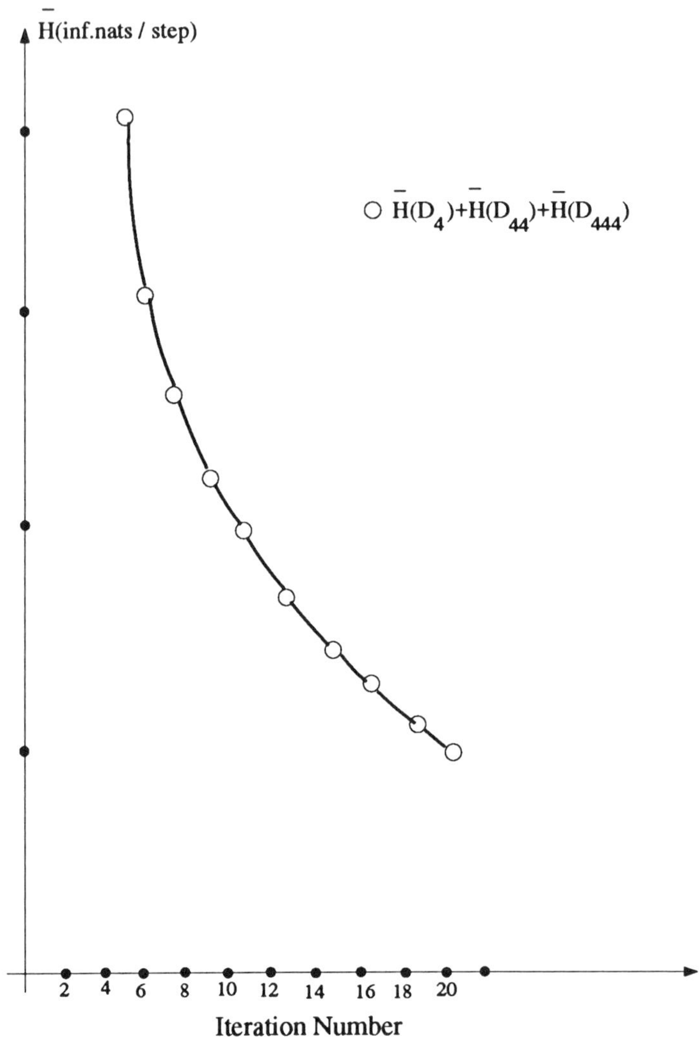

Figure 6.6 Evolution of Total Entropy Rate of Activated Memory Blocks for Sequential Testing of $c_4 = u_4$

The vision system is activated first to identify *SV1* and determine its position. Based on the feedback information, fl or f2, a path is chosen which is followed to operate on the valve and return to the initial position. A penalty function is also chosen. The plan is executed and the total cost is calculated. If the plan is executed successfully then the particular strings from which it has been formulated are rewarded, while the others are penalized. Upgrade of the probabilities and the information in the long term memory follows as explained in chapters 4 and 5.

The results shown in Figures 6.3 to 6.6 have been obtained for sequential testing of u_4. It is shown that the algorithms converge within twenty iterations.

Example 2

A sequential testing of the command $c_{19} = u_{19} =$ "Replace *FL3*" has been chosen as the second example. The approach that must be followed is the same with the one followed in the previous example. The formulation and execution of plans is more complicated in this case. The results obtained are shown in Figures 6.7 to 6.10.

Figure 6.11 indicates the evolution of the minimum overall accrued cost associated with the three types of commands.

6.3 An Intelligent Robotic System Operating in a Hospital Environment

A prototype system operating in a hospital environment has been developed and tested by simulation. An expert system cell has been designed for the robotic system organizer to incorporate the system functions. The purpose of this simulation is to demonstrate the feasibility of a robotic system assisting an immobile person in a hospital or house environment. The operation of the coordination level has been assumed to be known through a cost function related table.

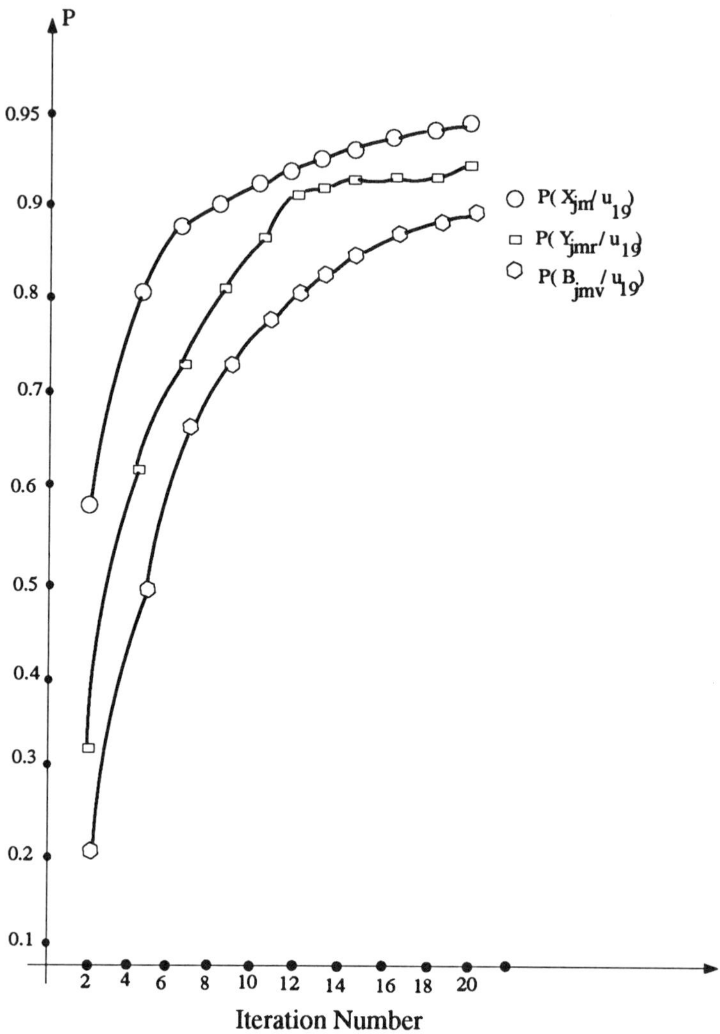

Figure 6.7 Evolution of Probability Distribution Function for Sequential Testing of $c_{19} = u_{19}$

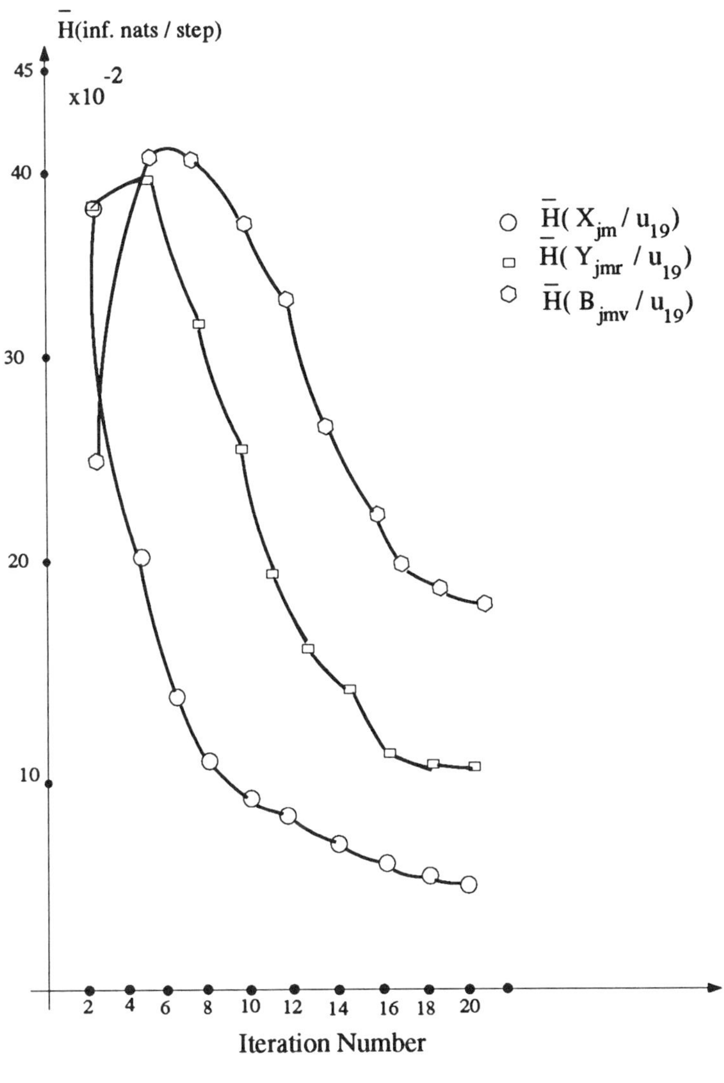

Figure 6.8 Evolution of Entropy Rates for Sequential Testing of $c_{19} = u_{19}$

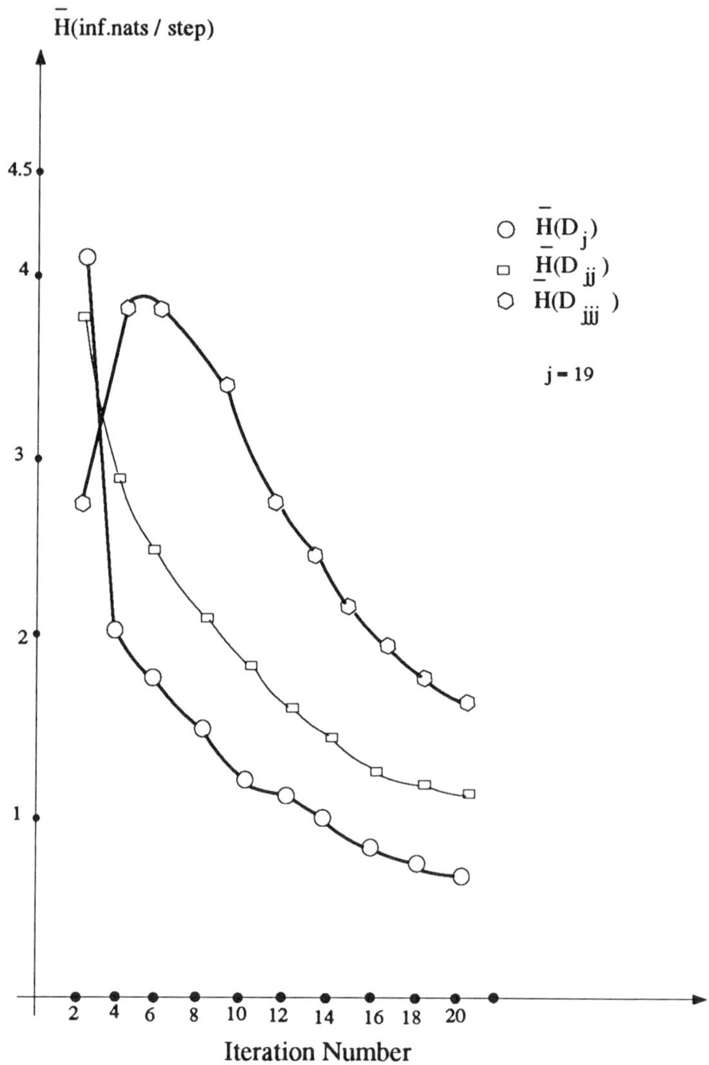

Figure 6.9 Evolution Rates of Activated Memory Blocks for Sequential Testing of $c_{19} = u_{19}$

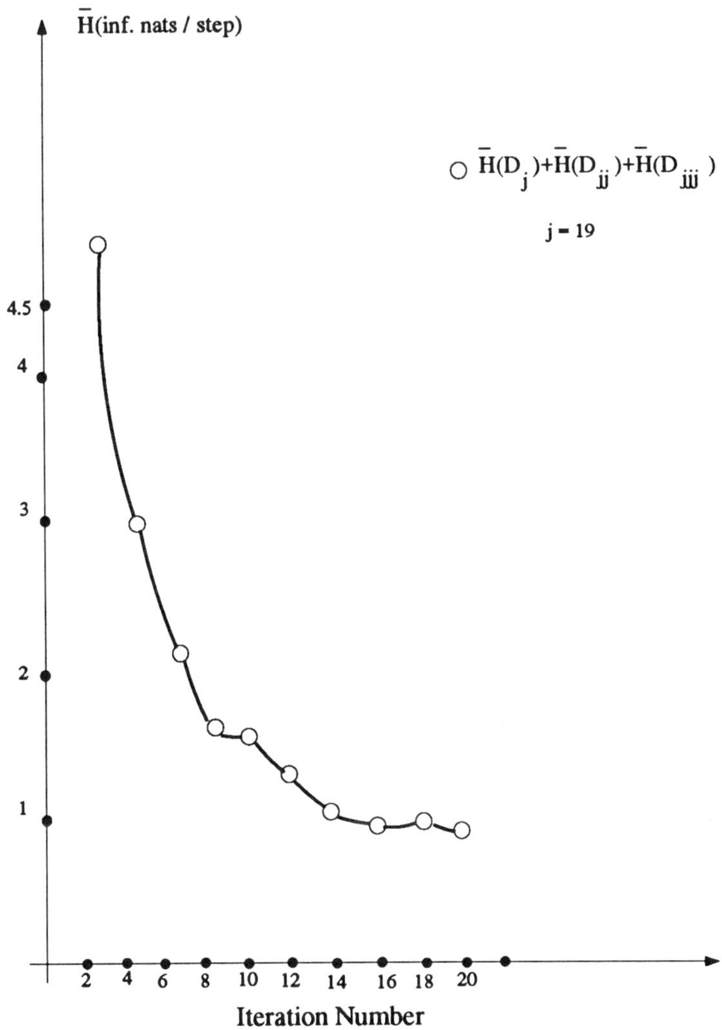

Figure 6.10 Evolution of Total Entropy Rate of Activated Memory Blocks for Sequential Testing of $c_{19} = u_{19}$

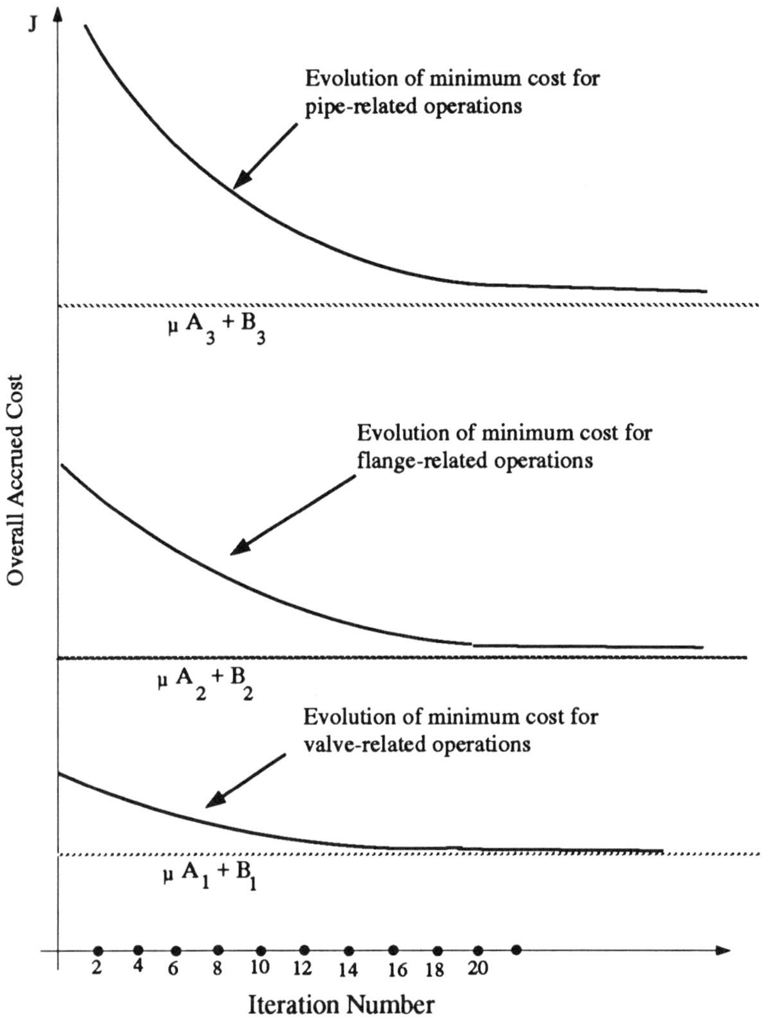

Figure 6.11 Evolution of the Minimum Overall Accrued Coordination Level Cost

Consider an immobile handicapped person within a room. Assume that the handicapped person is laying on a couch in the room which contains: ceiling light, desk lamp, radio, TV, stereo, book, briefcase, earphone, telephone, closet, door, and window; the person issues typed in commands to the robot, like: TURN-ON-TV, TURN-OFF-RADIO, GET-TELEPHONE, OPEN-WINDOW and CLOSE-DOOR. In this type of environment, we can define a priori a set of relevant basic task functions which are necessary to carry out the desired user commands. These basic task functions are listed according to the specific coordinator they are associated with, as shown in Figure 6.12 A list of possible user commands is also given in Figure 6.13. The general sequence of operations for each command type is also provided in Figure 6.14. Every command type is made up of a sequence of vision phase, motion planning phase, and action phase. The first two phases are identical for all of these three types of commands, because a logical progression of basic task functions is evoked in sequence to perform specific tasks. In essence, once a user command is issued, the robotic system first analyzes its environment and makes plans accordingly to carry out the desired tasks and execute the plan as accurately as possible. In the case when a task sequence does fail, then the feedback from the analysis of the plan execution will determine the reason, and consequently, what action the organizer needs to take next, in order to accomplish the requested job.

Various simulations based on different system parameters have been conducted. The overall performance analysis of the robotic system organizer in these simulations, is based on the necessary number of attempts the organizer needs to generate a correct task plan in order to carry out the desired user command. The system has been tested under many different conditions associated with:

1. The size and content of the existing knowledge base.
2. The existence of the particular command and/or command type within the knowledge domain, and,
3. The method used to generate the task plan.

I. Visual System Coordinator

1) IDENTIFY
2) LOCATE
3) SCAN

II. Sensory System Coordinator

4) INSENSE
5) EXSENSE

III. Motion System Cordinator

6) CHOOSE_PATH
7) COST_ANALYSIS
8) MOVE_BACK
9) MOVE_TO
10) PENALTY_FUNCTIONS
11) SELECT_D_MODEL
12) WC_TO_JC_CONVERSION

IV. Gripper System Coordinator

13) DUMP
14) FLIP
15) GRASP
16) GRIP
17) LIFT
18) PULL
19) PUSH
20) TURN

Figure 6.12 Basic Task Functions

I. TURN_ON / OFF

1) CEILING_LIGHT
2) DESK_LIGHT
3) RADIO
4) STEREO
5) TV

II. GET

6) BOOK
7) BRIEFCASE
8) CLOTHES
9) DRINKS
10) EARPHONE
11) TELEPHONE

III. OPEN / CLOSE

12) CLOSET
13) DOOR
14) WINDOW

Figure 6.13 Defined User Commands and Command Types

I. TURN_ON / OFF

1) SCAN
2) IDENTIFY
3) LOCATE
4) CHOOSE_PATH
5) SELECT_D_MODEL
6) MOVE_TO
7) GRIP
8) PUSH / FLIP
9) MOVE_BACK

II. GET

1) SCAN
2) IDENTIFY
3) LOCATE
4) CHOOSE_PATH
5) SELECT_D_MODEL
6) MOVE_TO
7) LIFT
8) MOVE_TO
9) DUMP
10) MOVE_BACK

III. OPEN / CLOSE

1) SCAN
2) IDENTIFY
3) LOCATE
4) CHOOSE_PATH
5) SELECT_D_MODEL
6) MOVE_TO
7) GRASP
8) TURN / LIFT
9) PULL / PUSH
10) MOVE_BACK

Figure 6.14 Operation Sequences for the Defined User Commands

The output of each simulation is a listing of the task plans that are generated, calculations of the probability of success of the best plan for each individual trial, and a graphical plot of these probabilities with respect to the trial number. These simulation outputs are shown in Figures 6.15 to 6.18 for four different input commands. A complete analysis of the results may be found in [235].

The system has the ability to analyze its own faults through evaluation mechanisms. This, in turn, allows the organizer to correct possible mistakes associated with the decision making and generate the best (alternate) possible task plan, if there exists one. The evaluation of the chosen task plan determines the following possible sources of errors:

1. Failure due to incorrect classification of the user command type.
2. Failure due to insufficient number of primitive events to generate the basic task functions.
3. Failure due to incorrect usage of basic task functions, and,
4. Failure due to non existing basic task functions.

Since each command type contains a certain function which is unique and is used only by that type of commands, if a failure is due to an incorrect command type, then the entire sequence must be regenerated and, thus, fall into another command type category. Sometimes the generated task plan may be incomplete due to the complicated nature of the task that is being issued. In this case, when detected, the organizer will try to fill in the necessary details that are needed to execute this particular command by inserting possibly additional repetitive basic task functions into the existing sequence. If the error is not created by either of the above, then an incorrect usage of the basic task function may be the cause. When this is the case, by identifying which of the task functions are incorrect, an alternate task sequence is generated by substituting the faulty ones with other task functions. Finally, if none of the above creates the failure (most likely), a new basic task function is needed, which however, does not currently exist in the knowledge base. Then, the organizer asks the user for help, and through teaching and/or addition of additional basic primitive events, the system becomes capable of generating the desired sequence.

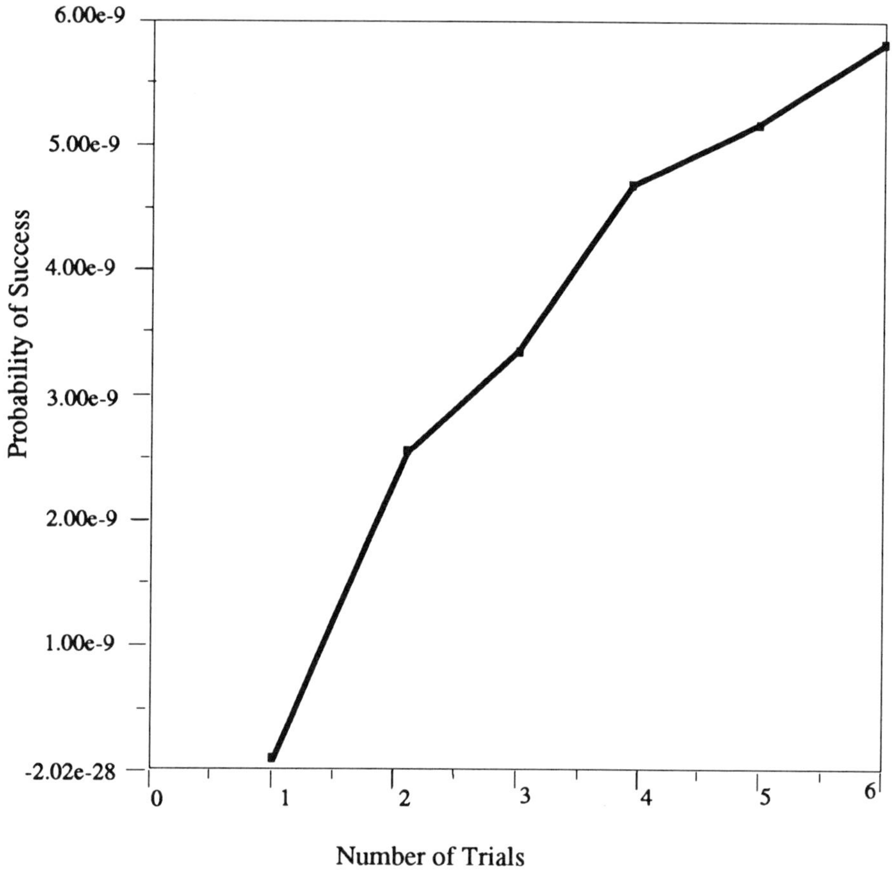

Figure 6.15 Simulation of Command: Get Clothes

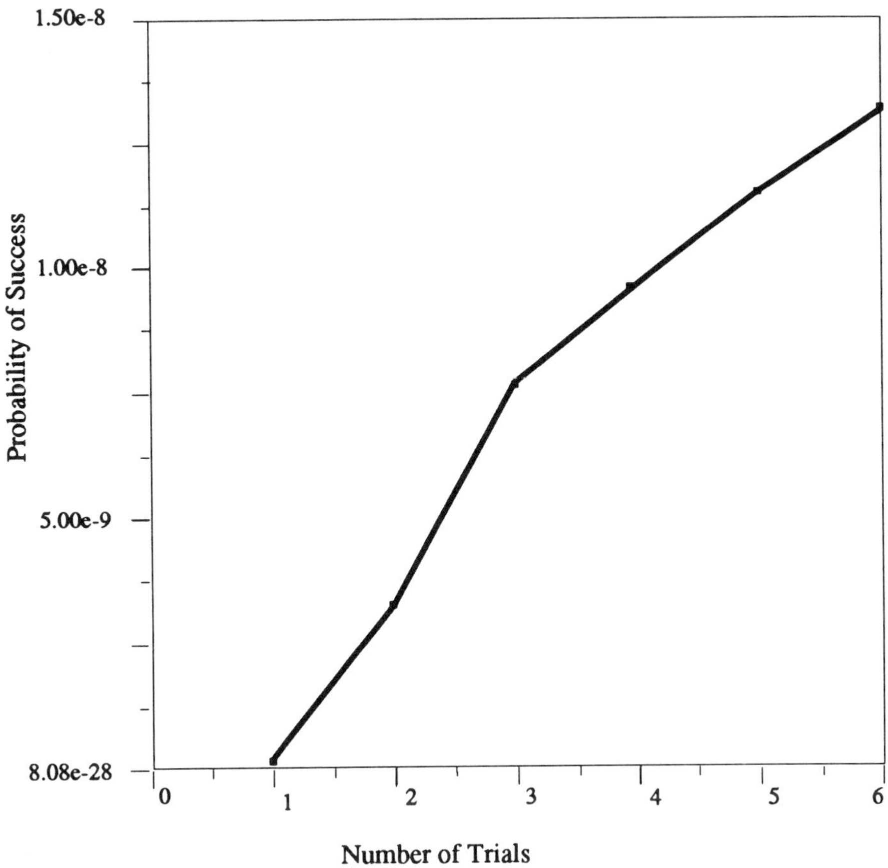

Figure 6.16 Simulation of Command: TURN_ON_TV

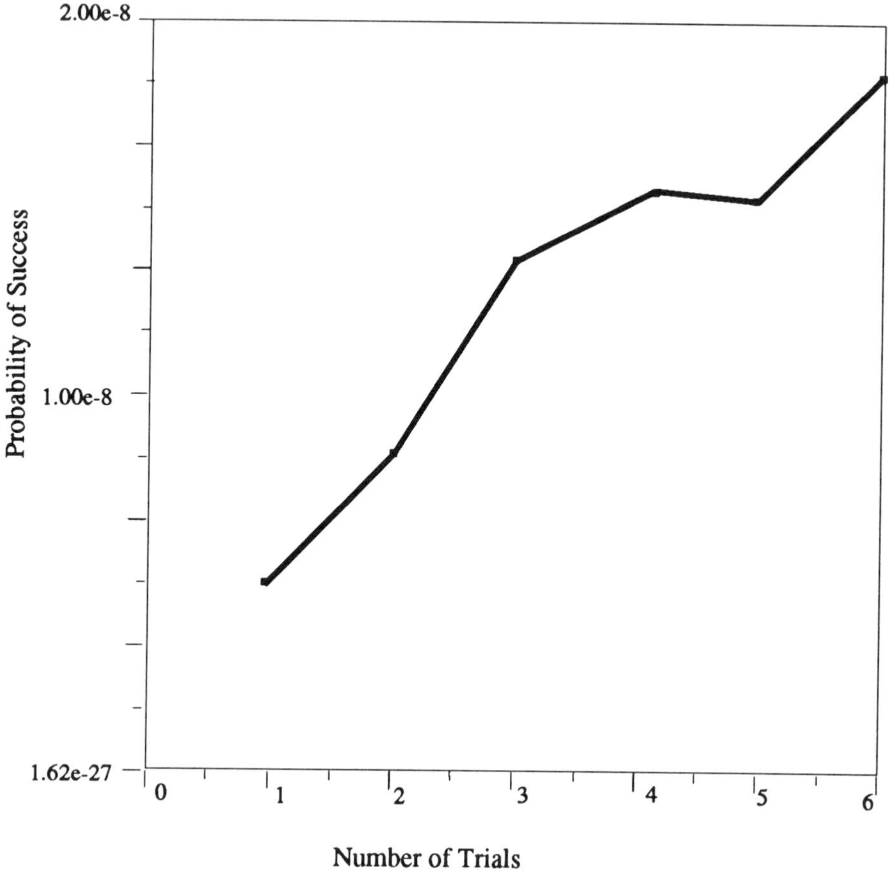

Figure 6.17 Simulation of Command: OPEN_CLOSET

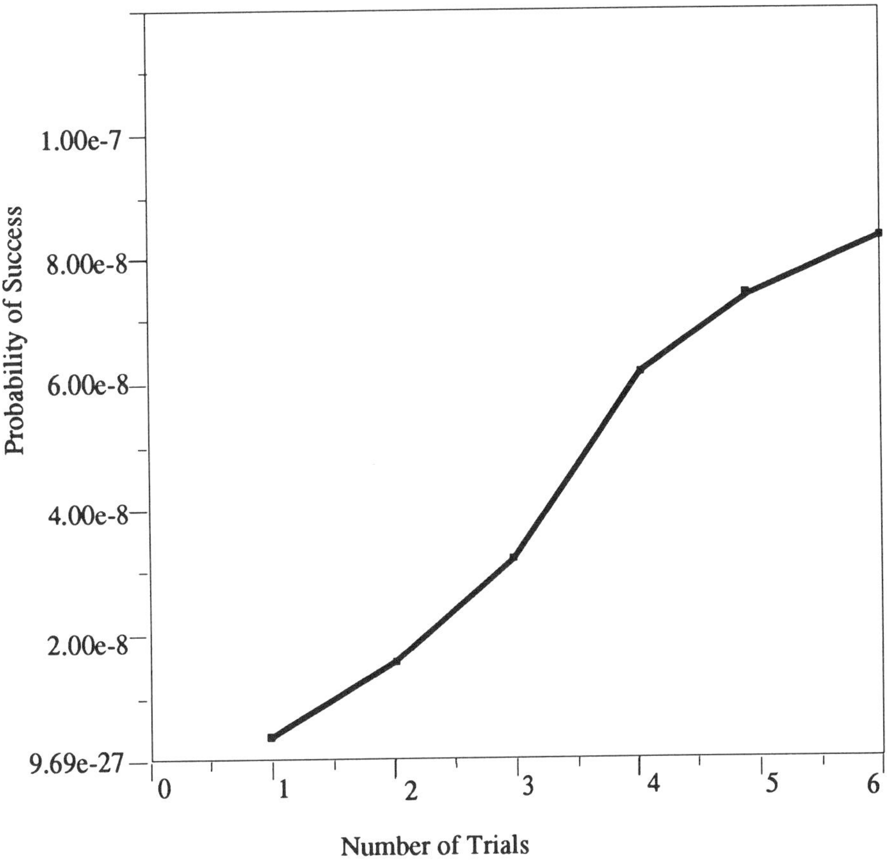

Figure 6.18 Simulation of Command: GET_BOOK

In conclusion, depending on the type of failure determined by the evaluation system, several more iterations may be necessary to formulate the correct task plan sequence. Other factors affecting the probability of success of generating the correct task plan include: how complicated the user command is, the size of the existing knowledge base, and the complexity of the knowledge domain. Assumptions and restrictions may affect the generation/formulation of the correct task sequence.

6.4 A Robotic Assembly System

The robot workstation in Figure 6.19 is designed to assemble a product from four different parts A, B, C and D, shown in Figure 6.20.

The product assembly is based on the following assumptions:

1. Parts A, B, C and D are fed from a conveyor belt at random order, but they always come together. When a part is at location F, the conveyor stops until the part is buffered or fetched; the next part is then brought to F.
2. The manipulator can execute the following actions:

    a. *buffering*— temporarily store parts located in the conveyor belt in fixed locations in the workstation.
    b. *mating*— joining of subassemblies (parts) one of which must already be held by the manipulator.
    c. *fetching*—retrieving a part in the parts buffer or directly from the conveyor belt; in the case of subassemblies, retrieving from designated locations in the workspace.

3. Parts A, B, C and D are buffered in locations B_1, B_2, B_3 and B_4, respectively. Part A always has to be buffered, while parts B, C and D may or may not be buffered.
4. There are two basic assembly sequences:

    a. Assuming A is already in buffer B_1, B is fetched either from buffer B_2 or directly form the conveyor belt and the mating operation involving B and A (resulting in subassembly S_1) is done in B_1. In a similar manner, if D is already in B_4, C is fetched either form the conveyor belt or from B_3, and the mating operation involving C and D (resulting in subassembly S_2) is done in B_4. Once S_1 and S_2 are available, S_2 is fetched from buffer B_4 and moved to buffer B_1 to be mated with S_1. S_1 and S_2 when mated together result in the final product S_4 which is then moved to G.

b. This sequence differs from the above in that after S_1 is assembled, C is fetched and moved to B_1 to be mated with S_1 (resulting in subassembly S_3), and then D is fetched and again moved to B_1 to be mated with S_3 to produce the final product S_4. In the first sequence D has to be buffered, while in the second case, it could be fetched directly from the conveyor belt.

5. Assembly of S_1 may be done before or after assembly of S_2.
6. It is desired that if A is already in B_1, once B is seen in F, B is fetched right away and the mating of B to A follows next. In a similar manner, if D is already in B_4 once C is seen in F, C is fetched right away and mated to D. This is to eliminate the unnecessary bufferings of B and C.

The efficiency of this assembly operation depends on the ability to handle incoming parts at random order.

Based on the assumptions made, there will be four buffering operations, five mating operations, four fetching operations, and one operation to move the finishing product to the second conveyor belt.

Although this example considers only one user command, it should be noted that the algorithms will likewise work for a finite number of user commands. An ideal classifier is considered.

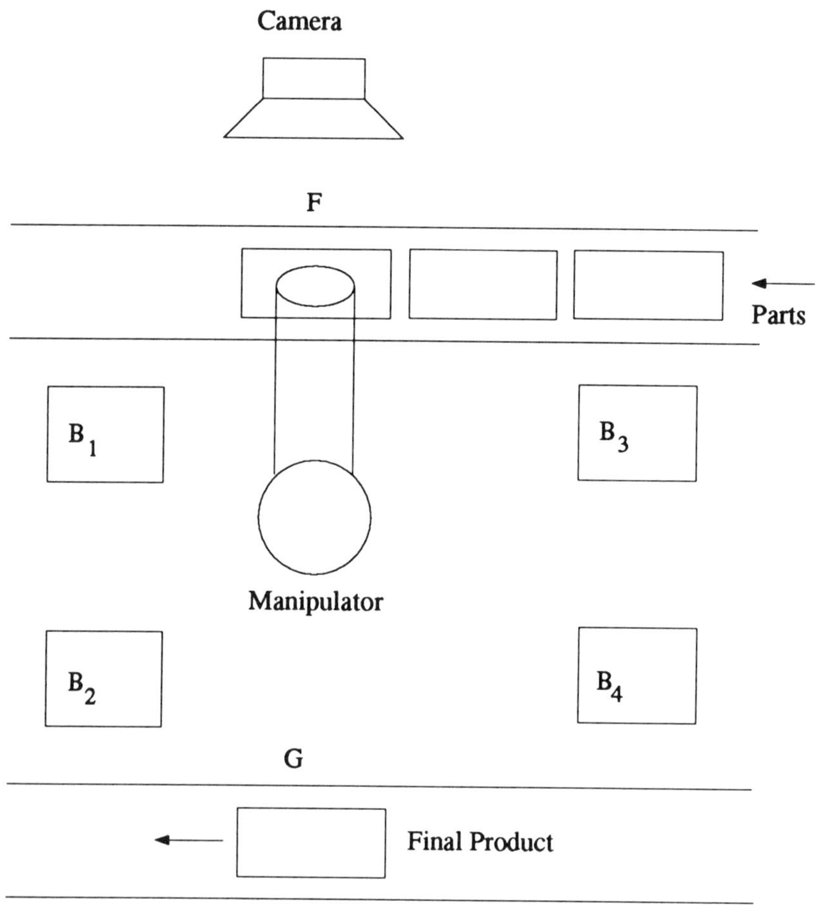

Figure 6.19 Robot Workstation

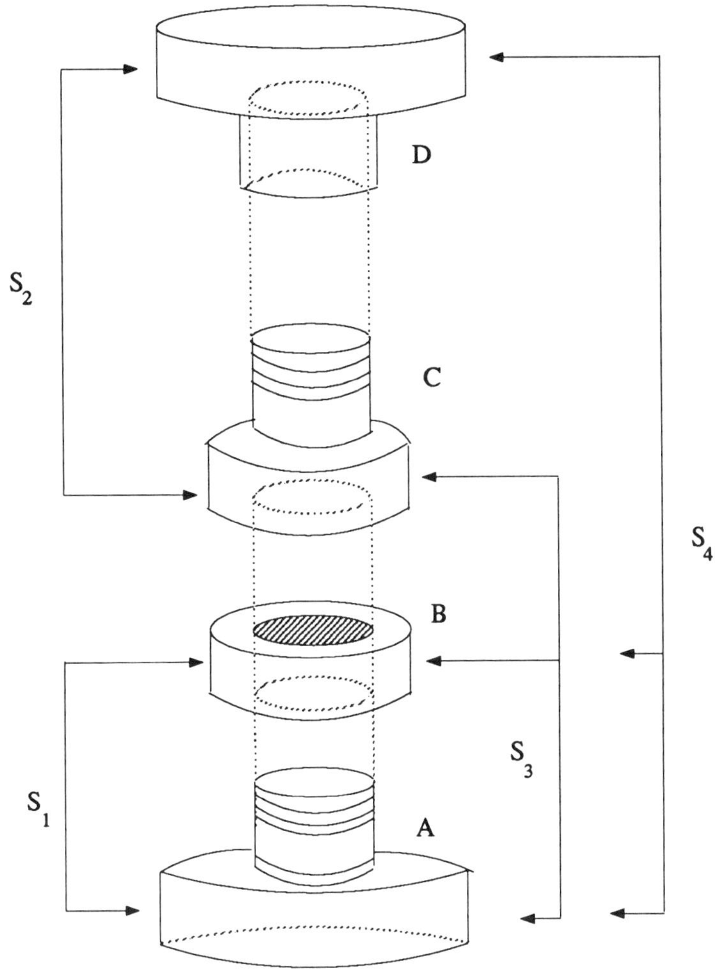

Figure 6.20 Description of Parts A, B, C and D and Sub-assemblies S_1, S_2 and S_3. The Final Product is S_4

The user command is: $c_1 = u_1 =$ assemble part.

The task domain is given below (*N=16, N-L=14, L=2*), along with the corresponding subsets from Table 1, chapter 4.

$$E_t = \{e_1, e_2, e_3, e_4, e_5, e_6, e_7, e_8, e_9, e_{10}, e_{11}, e_{12}, e_{13}, e_{14}, e_{15}, e_{16}\} \quad (6.4.4)$$

e_1 = buffer A

e_2 = buffer B

e_3 = buffer C

e_4 = buffer D

e_5 = fetch B

e_6 = fetch C

e_7 = fetch D

e_8 = fetch S_2

e_9 = mate C to D

e_{10} = mate B to A

e_{11} = mate C to S_1

e_{12} = mate D to S_3

e_{13} = mate S_2 to S_1

e_{14} = move S_4

e_{15} = see

e_{16} = move

$$
\begin{aligned}
E_{nr} &= \{e_1, e_2, e_3, e_4, e_5, e_6, e_7, e_8, e_9, e_{10}, e_{11}, e_{12}, e_{13}, e_{14}\} \\
E_r &= \{e_{15}, e_{16}\} \\
E_{sr} &= \{e_{15}\} \\
E_{end} &= \{e_{14}\} \\
E_{comp} &= \{(e_1\, e_3), (e_1\, e_4), (e_1\, e_5), (e_1\, e_6), (e_2\, e_1), \\
&\quad (e_2\, e_3), (e_2\, e_4), (e_2\, e_6), (e_3\, e_1), (e_3\, e_2), \\
&\quad (e_3\, e_4), (e_3\, e_5), (e_4\, e_1), (e_4\, e_2), (e_4\, e_5), \\
&\quad (e_4\, e_6), (e_5\, e_{10}), (e_6\, e_9), (e_6\, e_{11}), (e_7\, e_{12}), \\
&\quad (e_8\, e_{13}), (e_9\, e_1), (e_9\, e_2), (e_9\, e_5), (e_9\, e_8), \\
&\quad (e_{10}\, e_3), (e_{10}\, e_{14}), (e_{10}\, e_6), (e_{10}\, e_8), (e_{11}\, e_7), \\
&\quad (e_{16}\, e_1), (e_{16}\, e_2), (e_{16}\, e_3), (e_{16}\, e_4), (e_{16}\, e_5), \\
&\quad (e_{16}\, e_6), (e_{16}\, e_7), (e_{16}\, e_8), (e_{16}\, e_9), (e_{16}\, e_{10}), \\
&\quad (e_{16}\, e_{11}), (e_{16}\, e_{12}), (e_{16}\, e_{13}), (e_{16}\, e_{14}), (e_1\, e_{15}), \\
&\quad (e_2\, e_{15}), (e_3\, e_{15}), (e_4\, e_{15}), (e_5\, e_{15}), (e_6\, e_{15}), \\
&\quad (e_7\, e_{15}), (e_8\, e_{15}), (e_9\, e_{15}), (e_{10}\, e_{15}), \\
&\quad (e_{11}\, e_{15}), (e_{12}\, e_{15}), (e_{13}\, e_{15}), (e_{12}\, e_{14}), \\
&\quad (e_{13}\, e_{14})\}
\end{aligned}
\tag{6.4.5}
$$

The organization level algorithm follows:

The original set of pertinent activities associated with the input command is $(2^{16} - 1)$. According to the operational procedures, the following justifiable simplifications are considered:

It can be verified that $(e_{15}\, e_{16})$ is the only valid repetitive ordering. Store $E_{ord} = \{(e_{15}\, e_{16})\}$ in the long term memory of the organizer. Then, given the command c_1, we have:

$$
\begin{aligned}
E_c &= \{e_1, e_2, e_3, e_4, e_5, e_6, e_7, e_8, e_9, e_{10}, e_{11}, e_{12}, e_{13}, e_{14}\} \\
E_{cr} &= \{\{e_1, e_4, e_5, e_6, e_8, e_9, e_{10}, e_{13}, e_{14}\}, \\
&\quad \{e_1, e_5, e_6, e_7, e_{10}, e_{11}, e_{12}, e_{14},\}\},\ q = 8 \\
E_{snr} &= \{e_1, e_2, e_3, e_4\},
\end{aligned}
$$

$$
\begin{aligned}
E_{ord} = & \ \{(e_{15}e_{16})\}, \\
E_{sr} \ = & \ \{e_{15}\}, \\
(e_i e_{15}) & \in E_{comp}, \ \forall i \in \{1.....13\} \\
(e_{16} e_j) & \in E_{comp}, \ \forall i \in \{1.....14\}
\end{aligned}
\tag{6.4.6}
$$

To demonstrate the feasibility of the algorithm and the necessity of the operational procedures, several different cases were considered in [236], related to the complete presence/absence of the precedence constraints, incompletely specified precedence constraints, and incompletely specified critical events. All cases demonstrated how the *precedence constraints make the algorithm robust by preventing generation of absurd activities and generation of meaningful activities that do not satisfy the imposed assumptions.*

It has been shown in [236] that a set of 76 activities represents all possible valid sequences of active non repetitive events which also satisfy all assumptions made. It follows that for each activity (with non repetitive primitives) of the form $(e_i, e_j, ..., e_k)$, the corresponding plan is $(e_{15}\, e_{16}\, e_i\, e_{15}\, e_{16}\, e_j\, e_{15}\, e_{16}\, ...e_{15}\, e_{16}\, e_k)$.

To demonstrate the convergence of the algorithm, a table look up cost of execution of primitive tasks has been considered, along with the following assumptions: The cost of execution of the repetitive events is negligible compared to the cost of execution of the non repetitive ones. Therefore, the cost of execution of each plan is proportional to the cost of the corresponding activity of critical non repetitive events. The total cost is assumed to be the sum of the individual costs. Moreover, it is considered that it is possible to identify the order parts appear in F.

Based on the above assumptions and from the fact that for this example, there is only one plan associated with any activity it follows that plans and activities may be used interchangeable without loss of generality.

Table 2 lists the estimated and actual costs for executing the non repetitive events. Normally, the actual costs of plans are determined from the data fed back from the coordinators. From the assumption that parts come at random order, it follows that there are 24 ways (4!) in which parts A, B, C and D could come and be seen at location F.

Table 2: Actual and Estimated Costs of the Non Repetitive Events

Event	**Estimated Cost**	**Actual Cost**
e_1	2	6
e_2	2	3
e_3	2	2
e_4	2	4
e_5	3	5
e_6	3	4
e_7	3	5
e_8	4	7
e_9	6	8
e_{10}	5	9
e_{11}	7	8
e_{12}	8	9
e_{13}	9	11
e_{14}	5	7

Given a part sequence, the plan chosen is the one with the highest probability among the plans that are relevant to that sequence as previously explained. After execution of a plan, the costs of the plans are recomputed in the following manner; initially, the costs of plans are determined from the cost estimates alone; subsequently, the actual costs are used for events that have been used in previous executions.

The plan which has the minimum cost among those which are relevant to the part order is rewarded, resulting in an increase of its probability. The rest are penalized with consequent decreases in corresponding probabilities.

Three plans were chosen to demonstrate the convergence of the learning algorithm. There are 40 training sessions (iterations). For the first 20 training sessions, parts are presented in order ACDB (plan 2); during the next 10, ADCB (plan 3) and during the last 10, ABDC (plan 13).

Figure 6.21 shows the plots of probabilities for 40 iterations. Initially, the plans are assigned equal probabilities. During the first 20 iterations, the probability of Plan 2 is increasing because it is constantly rewarded. Plan 2 has the lowest cost of execution of 3 potential candidates. During the next 10 iterations, the probability of plan 9 is increasing while that of plan 2 is decreasing. This is because only this plan is relevant to parts order ADBC. Hence, it is constantly rewarded. During the last 10 iterations, it has been shown that plan 13 is the one with the lowest cost. Since plan 13 has the lower cost of the two, it is constantly being rewarded, hence a resulting increase in its probability [236]. The probabilities of all the plans except plans 2, 9, and 13 are monotonically decreasing and are negligible.

6.5 Remarks

The example have demonstrated the applicability of the models presented in chapter 4. However, in real life situations, all cost are computed in real time and the coordination and execution levels function in real time. When this is the case, timing requirements and penalty functions are very important. This is an open topic of research.

The architectural models are easily applied to all three examples.

The last chapter summarizes, concludes this monograph and also reviews briefly very recent advances in the area of Intelligent Robotic Systems.

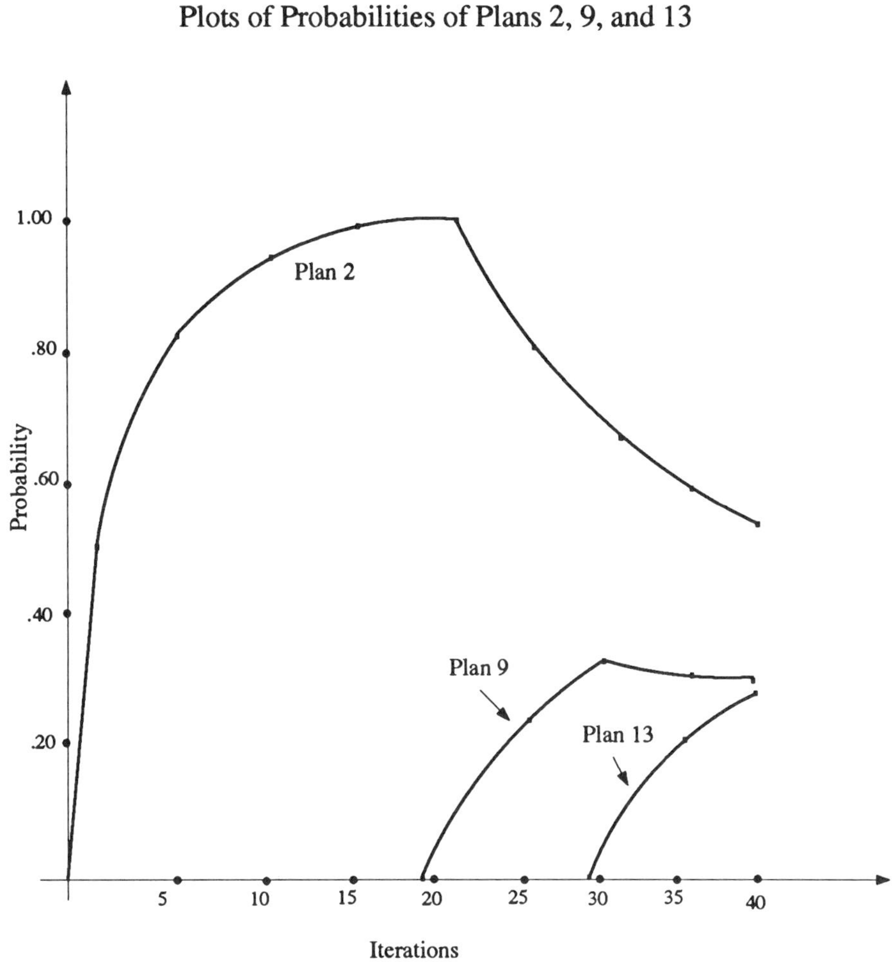

Figure 6.21 Evolution of the Probability of Success of the Best Plans

Chapter 7
CONCLUSIONS AND RECOMMENDATIONS

7.1 Summary

This monograph has concentrated on the development of a unified mathematical foundation for the analytical design of Intelligent Robotic Systems operating in uncertain environments.

Three factors have motivated our research and have contributed to the completion of this monograph:

1. Intelligent Machines may be driven by controls with special characteristics described by the methods of hierarchically intelligent control systems. The same methods apply to Intelligent Robotic Systems.

2. Intelligent control systems have self-organizing capabilities and, therefore, Intelligent Robotic Systems may operate with minimum supervision and interaction with a human operator, and,
3. The entropy function as defined in Information Theory and Theoretical Thermodynamics has been proven to be a sufficient analytic measure that unifies the performance of the three interactive levels of such systems, and provides a means to describe quantitatively the flow of knowledge (information) within such machines and systems.

The modeling approach presented, consists of three interactive levels incorporating multi layer and multi level control. The system structure is designed according to the principle of Increasing Precision with Decreasing Intelligence.

The organization level has been modeled after a knowledge based system with functions: machine reasoning, machine planning, machine decision making, machine feedback and long term memory exchange, by interpreting mathematically concepts and ideas used in knowledge based systems of Artificial Intelligence. A probabilistic algorithm and an expert system cell are derived to accommodate the organizer functions.

The model of the coordination level is composed of a number of coordinators. The coordinators are of fixed structure and do not communicate with each other. A dispatcher of variable structure serves as the communicator between the different coordinators. Each of the individual coordinators performs its own prespecified and fixed individual functions. Specific hardware (execution devices) are associated with each coordinator which execute specific tasks when a command is issued by its corresponding coordinator. An algorithm has been derived for the coordination level functions.

The feedback mechanisms for the three interactive levels are explicitly derived. Two types of feedback information are defined: on line feedback or real time feedback and off line feedback. On line feedback information is communicated (in terms of individual cost functions) from the execution level to the coordination level during the completion of a user requested job. Off line feedback information is communicated (in terms of an accrued cost associated with the overall execution of the job) from the coordination level to the organization level after the completion of a user requested job. An integrated control and diagnostics mechanism is proposed to accommodate payload variations and thus, overcome this type of potential local failure.

A major advantage of the mathematical modeling is that they are very general and easily modifiable to account for more, or less operational constraints based on the specific application problem. Moreover, an advantage of the architectural models derived, is that their structure is suitable not only for this mathematical approach but also for linguistic based design of Intelligent Robotic Systems or Fuzzy Set based design.

7.2 Present and Future Developments in Intelligent Controls

In the last few years, Intelligent Machines have reached a point of maturity to be implemented on a robotic testbed aimed for space assembly and satellite maintenance. They also feature an application of the Hierarchically Intelligent Control Theory which is based on the principle of Increasing Precision with Decreasing Intelligence to form an analytic methodology using entropy as a measure of performance. The three levels, organization, coordination and execution, representing the original architecture of the system, have not been changed, but their internal architectures have been recently modified to incorporate more efficient and effective structures dictated by experience.

The Organization Level

A *Boltzmann machine* type neural net, originally proposed for text generation, has been used for the structure that implements the organization level of Intelligent Machines [148, 199]. This machine would connect a *finite* number of letters (nodes) into grammatically correct words (rules), by minimizing at the first layer the total entropy of connections. Replacing the letters at the nodes with words, at second layer, sentences are created. At the third layer, the words are replaced by sentences at the nodes, etc., until a meaningful text is created.

The functions of the organizer, following, again, the model of a general knowledge based system, comprise of: *representation, abstract task planning, decision making, learning* (from experience). At this level, abstract task planning occurs with minimum knowledge of the current environment. All those functions may be generated by a Boltzmann machine similar to the text generating machine, by considering a finite number of primitive elements at the nodes, constituting the basic actions and actors at the *representation phase.* Strings of these primitives are generated by the Boltzmann machine at the *planning phase* with the total entropy representing the cost of connections. The selection of the string with minimum entropy is the *decision making* process. The upgrading of the parameters of the system by rewarding the successful outcomes through feedback, is the *learning* procedure. The next to minimum entropy string may be retained as an alternate plan in case of failure of the original, or errors created by the environment.

This bottom-up approach, characteristic of natural languages, is extremely simple and effective, utilizing intelligence to replace the complexity of the top-down type task decompositions. The tasks thus generated, are practically independent of the current environment. Information about the present world should be gathered at the coordination level. An appropriate world model is constructed from sensory and motion information available at that level. However, the structure of the dispatcher, designed to interpret the organizer's strings, monitor and traffic commands among the other coordinators, is highly dependent on the strings which represent the planned tasks.

Petri net Transducers (PNT) have been used to model the coordinators at the next level. A Petri net generator is required to create the dispatcher's PNT for every task planned. This may be accomplished by another Boltzmann machine or a part of the existing plan generating architecture.

The Coordination Level

The coordination level is a tree structure of PNTs as coordinators, with the dispatcher as the root [249–254]. These devices set up the communication protocols with the help of small real time memories and serve apply in real time the rules generated by the organization level to properly generate and sequence the subtasks in the particular environment of the machine, in order to execute the given original command. The PNT for the dispatcher is generated by the organizer for every specific plan and is transmitted, asynchronously, to the coordination level along with the plan to be executed. The function of the dispatcher is to interpret the plan and assign individual tasks to the other coordinators, monitor their operation, and transmit messages and commands from one coordinator to another as needed. As an example, a command is sent to the vision and sensing coordinator to generate a model of the environment, the coordinates of the objects for manipulation to be tabulated, and then transmitted to the motion coordinator for navigation and motion control. This command is executed by having each transition of the associated Petri nets to initialize a package corresponding to a specific action. These packages are stored in short memories associated with each of the coordinators.

The rest of the coordinators have a fixed structure with alternate menues available at request. They communicate commands and messages with each other, through the dispatcher. They also provide information about reception of a message, data memory location and job completion.

No data is communicated at the coordination level, since the task planning and monitoring may be located in a remote station, and such an exchange may cause a channel congestion. A preferred configuration for such situations is that the coordinators with a local dispatcher may be located with the hardware at the work site, while a remote dispatcher, connected to the organizer, interacts with the local one from a remote position. This concept simplifies the communication problem considerably, since only short messages are transmitted back and forth through a major channel between local and remote stations, requiring a narrow bandwidth.

Entropy measures have been developed for each coordinator such that they may be used to minimize the complexity and improve the reliability of the system.

The Execution Level

The execution level contains all the hardware required by the Intelligent Machine to execute a task. There is a one to one correspondence between hardware groups and coordinators. Therefore, their structure is usually fixed. This level also contains all the drivers, VME buses, short and fast memory units, processors, actuators and special purpose devices needed for the execution of a task [202]. After the successful completion of a job, feedback information is generated at this level for evaluation and parameter updating of the whole machine. Complexity dominates the performance of this level. Since *precision* is proportional to *complexity*, it also defines the amount of effort required to execute a task. It has been shown that all the activities of this level may be measured by entropy, which may serve as a measure of complexity as well. Minimization of local complexity through feedback, may serve as local design procedure.

Entropy Formulation of Motion Control: The cost of control at the hardware level may be expressed as an entropy which measures the uncertainty of selecting an appropriate control to execute a task. By selecting an optimal control, one minimizes the entropy, the uncertainty of execution. The entropy may be viewed in that respect as an energy in the original sense of Boltzmann, satisfying Jaynes' principle of maximum entropy. The result is a theory of optimal control based on the minimization of ensuing entropy.

Entropy measure of the Vision System: The Optimal Control Theory, designed mainly for motion control, may be implemented for vision control, path planning and other sensory systems pertinent to an Intelligent Machine by slightly modifying the system equations and cost functions. After all, one is dealing with real time dynamic systems which may be modeled by a set of dynamic equations. To be more specific, consider a stereo vision system of a pair of cameras mounted at the end of a robot arm. The system may be positioned at different viewpoints to reduce problems with noise, considered one at a time due to time limitations. The accuracy of measuring the object's position depends upon its relative position in the camera frame. Consequently, each viewpoint will have different measurement error and time statistics. These statistics generate the appropriate entropies that define the uncertainty of the measurement of the vision system. These entropies are minimized to obtain optimal results [131].

The architecture necessary to accommodate the revised system functions, does not differ substantially from the original architecture described in the previous chapters. Some details are more elaborate and more efficient. The main contribution is a more efficient structure, effective, versatile and capable of remote operation.

7.3 Some Ideas for Future Investigation

Further research is needed to consolidate the results of this monograph. Several suggestions for possible consideration in the research for Intelligent Machines are considered in the sequel.

1. A Fuzzy Set formulation of the problem for the analytical design of Intelligent Machines will provide a generalized (and global) methodology for the organizer functions. The random variables x may be distributed over the interval [0,1] to reflect a membership function.
2. Design of Intelligent Machines with judgement capabilities will provide greater flexibility to plans organization, formulation and execution. In this case, the Mathematical Theory of Evidence is a sufficient mathematical tool for the design of such machines with judgement [206].

3. A linguistic based approach for the design of Intelligent Machines is an alternative solution. A comparison with the mathematical approach will provide useful information about possible deficiencies related to both of them.
4. A sensitivity analysis due to misinterpretation of the user commands will provide useful information about the noise effects to the Intelligent Machines.
5. An extension of the proposed mathematical approach to account for unpredictable events during plan execution is of great importance. When this is the case, the Intelligent Machine will be able to react to such situations and take alternative actions to complete the requested job.
6. Modifications of the mathematical and architectural models to account for interactive functions of the different coordinators may be of interest, and,
7. A reliability analysis of the individual functions of the Intelligent Machines will provide a measure of robustness of the machine and help in the selection of alternate programs.
8. More research work is necessary to optimize the Boltzmann machine and PNT based design.
9. A complete integrated scheme for simultaneous control and diagnostics will enhance further the capabilities of the system.

Appendix A REFERENCES

1. Acar, L. and U. Ozguner, "Design of Hierarchically Distributed Expert Controllers for Large-scale Systems." In *Proceedings of the 4th IEEE International Symposium on Intelligent Controls,* Albany, NY, September 1989.
2. Adiv, G., "Determining 3–D Motion and Surface Structure from Optical Flow Generated by Several Objects." *IEEE Transactions on Pattern Analysis and Machine Intelligence,* PAMI-7 (No. 6): 384–401, June 1985.
3. Albus, J.S., "Outline for a Theory of Intelligence." *IEEE Transactions on Systems, Man and Cybernetics*, Vol. SMC-21, No. 3, May/June 1991.
4. Albus, J. S., "A Theory of Intelligent Systems." *Proceedings of the IEEE International Symposium on Intelligent Controls*, 1990, Philadelphia, PA.
5. Albus, J. S., "The Role of World Modeling and Valve Judgment in Perception." *Proceedings of the IEEE International Symposium on Intelligent Controls*, 1990, Philadelphia, PA.
6. Albus, J. S., "*Brains, Behavior, and Robotics*," Byte/McGraw-Hill, Peterborough, NH, 1981.
7. Albus, J. S., "Hierarchical Interaction Between Sensory Processing and World Modeling in Intelligent Systems." *Proceedings of the IEEE International Symposium on Intelligent Controls*, 1990, Philadelphia, PA.
8. Albus, J. S., H. G. McCain and R. Lumia, "NASA/NBS Standard Reference Model for Telerobot Control System Architecture (NASREM)." *NIST, Technical Report 1235*, Gaithersburg, MD 1989.
9. Albus, J. S., J. A. Simpson, R. J. Hocken, "The Automated Manufacturing Research Facility of the NBS." *Journal of Manufacturing Systems,* (No. 1): 17–32, 1982.

10. Albus, J. S., "A New Approach to Manipulation Control: The Cerebellar Model Articulation Controller." *Transactions of ASME, J. Dynamics Systems, Measurement and Control,* 97, 220–227, 1975.
11. Alexander, E. M., B. Frederick, R. M. Carrell and G. J. Van Sant, "Handling of Irregular Sized Mailpieces by Adaptive Robotics." *IEEE Control Systems Magazine,* February 1989.
12. Andreadakis, S. K., *Analysis and Synthesis of Decision Making Organizations.* Ph.D. Thesis, LIDS-TH-1740, MIT, Cambridge, MA, 1988.
13. Antsaklis, P. J., K. M. Passino, S. J. Wang, "Towards Intelligent Autonomous Control Systems, Architecture and Fundamental Issues. " *Journal of Intelligent and Robotic Systems,* Vol. 1, No. 4, 1989.
14. Antsaklis, P.J. and K. M. Passino, "Artificial Intelligence Planning Problems in a Petri Net Framework." In *Proceedings of the 1988 American Control Conference,* Atlanta, GA, 1988.
15. Antsaklis, P.J. and K. M. Passino, "Planning Via Heuristic Search in a Petri Net Framework." In *Proceedings of the 3rd IEEE International Symposium on Intelligent Controls,* Arlington, VA, 1988.
16. Antsaklis, P. J. and K. M. Passino, "A System and Control Theoretic Perspective On Artificial Intelligence Planning Systems." *Applied Artificial Intelligence,* Vol. 3, 1989.
17. Ashby, W. R., "Two Tables of Identities Governing Information Flows Within Large Systems." *Communications of American Society of Cybernetics,* Vol. 1, No. 2, 2-7, 1969.
18. Astrom, K. J., "Expert Control." *Automatica,* Vol. 22, 1986.
19. Athans, M. and Falb, P., *Optimal Control,* McGraw Hill, New York, 1965.
20. Athans, M., "Command and Control Theory: A Challenge to Control Science." *IEEE Transactions on Automatic Control,* 32 No. 4, 286–293, April 1987.

21. Babcock S. M. and Barhen J., ''Real Time Algorithms for Robotic Control of Teleoperators.'' In *Proceedings of Robotic International Sponsored ''Robots 8'' Conference and Exposition,* June 4-7, Detroit MI.

22. Babcock S. M., Hamel W. R. and Martin H. C., ''Real-Time Control of Mobile Robot with Multiple Manipulators.'' *ORNL CESAR/TRP-05*, August 1983.

23. Ballard, D. H. and Brown, C. M., *Computer Vision*, Prentice Hall, 1982.

24. Banerji, R., *Artificial Intelligence: A Theoretical Approach,* North Holland, NY, 1980.

25. Barhen, J., Saussure , G. and Weisbin, C. R., ''Strategy Planning by an Intelligent Machine.'' *ORNL CESAR/TRP-01*, August 1983.

26. Barhen, J. and Hamel, W. R., ''Sensor Fusion for Intelligent Control.'' *ORNL CESAR/TRP-04,* August 1983.

27. Bayes, T., "An Essay Toward Solving A Problem in the Doctrine of Chances.'' *Philosophical Transactions of the Royal Society*, 53, 370-418, 1763.

28. Besl, P. J., "Active, Optical Range Imaging Sensors." *International Journal of Machine Vision and Applications,* 1(2): 127–152, 1988.

29. Besl, P. J., *Surfaces in Range Image Understanding,* Springer Verlag, New York, 1988.

30. Besl, P. J. and R. C. Jain, "Three Dimensional Object Recognition." *ACM Computing Surveys,* Vol. 17 (No. 1): 75–145, March 1985.

31. Birk, J.R. and R. B. Kelley, "An Overview of the Basic Research Needed to Advance the State of Knowledge in Robotics." *IEEE Transactions on Systems, Man and Cybernetics,* Vol. SMC-32, No. 8, 575–579, 1981.

32. Boettcher, K. L., "*An Information Theoretic Model of the Decision Maker*," M. S. Thesis, LIDS-TH-1096, MIT, Cambridge, MA, June 1981.
33. Boettcher, K. L. and A. H. Levis, "Modeling the Interacting Decision Maker With Bounded Rationality." *IEEE Transactions on Systems, Man and Cybernetics,* Vol. SMC-12, No. 3, May-June 1983.
34. Boettcher, K. L. and A. H. Levis, "On Modeling Teams of Interacting Decision Makers with Bounded Rationality." *LIDS-P-1189, MIT*, Cambridge, MA, March 1982.
35. Bolles, R. C. and R. A. Cain, "Recognizing and Locating Partially Visible Objects." *International Journal of Robotics Research,* Vol. 1, No. 3: 57–82, 1982.
36. Boltzmann, L., *Populari Schriften*, Leipzig, 1905.
37. Bonissone, P. P. and Valavanis, K. P., "A Comparative Study of Different Approaches to Qualitative Physics Theories." In *Proceedings of the 2nd Conference on Articicial Intelligence Applications,* December 1985, Florida, and, *General Electric Report No. 85CRD211.*
38. Broekstra, G., "On the Representation and Identification of Structure Systems." *International Journal of General Systems (9)* (1978), 1271-1293.
39. Brooks, R. A., "Symbolic Error Analysis and Robot Planning." *International Journal of Robotics Research,* Vol. 1, Number 4, Winter 1982.
40. Buchanan, B. G. and E. H. Shortliffe, *Rule-Based Expert Systems*, Addison Wesley, 1984.
41. Cavallo, R. E. and Klir, G. J., "Reconstructibility Analysis: Overview and Bibliography." *International Journal of General Systems* 1981, Vol. 7, 1-6.
42. Chapman, D., "Planning for Conjunctive Goal." *Technical Report TR:83–85, MIT, AI Lab*, Cambridge, MA, 1985.

43. Conant, R. C., ''Information Flows in Hierarchical Systems.'' *IEEE Transactions on Systems, Man and Cybernetics,* Vol. SMC–6, No. 4, April 1976.

44. Conant, R. C., ''Laws of Information which Govern Systems.'' *IEEE Transactions on Systems, Man and Cybernetics,* Vol. SMC-6, No. 4, April 1976.

45. Conant, R. C., ''Detecting Subsystems of a Complex System.'' *IEEE Transactions on Systems, Man, and Cybernetics,* September 1972, 550-553.

46. Conant, R. C., ''Detection and Analysis of Dependency Structures.'' *International Journal of General Systems*, 1981, Vol. 7, 81-91.

47. "Conceptual Information Model for an Advanced Factory Management System, Factory and Job Floor Level." *Technical Report, Final Report 84–FM-03, CAM-I Inc.*, August 1984.

48. Cypher, R. and J. L. C. Sanz, "SIMD Architectures and Algorighms for Image Processing and Computer Vision." *IEEE Transactions on Acoustics Speech and Signal Processing,* ASSP 37 (12): 2158–2174.

49. Decker, K. S., "Distributed Problem Solving Techniques, A Survey." *IEEE Transactions on Systems, Man and Cybernetics,* SMC-17, No. 5, 729–40, September-October 1987.

50. DeMello, L. H., Sanderson, A. C., "Two Criteria for the Selection of Assembly Plans: Maximizing the Flexibility of Sequencing the Assembly Tasks and Minimizing the Assembly Time Through Parallel Execution of Assembly Tasks." *IEEE Transactions on Robotics and Automation*, Vol 7., No. 5, 1991.

51. DeMello, L.H. and A. C. Sanderson, "And/Or Representation of Assembly Plans." In *Proceedings of the AIII-86,* Philadelphia, August 1986.

52. DeMello, L.H., and A. C. Sanderson, "Real-Time Planning and Intelligent Control." *Technical Report CIRSSE Document No., 31, RPI,*, 1988.

53. DeMello, L. H., *Task Sequence Planning for Robotic Assembly,* Ph.D. Thesis, Carnegie Mellon University, 1989.

54. Drenick, R. F., ''Organization and Control.'' In *Directions in Large Scale Systems,* Ho and Mitter, Eds., Plenum NY, 1976.

55. Drenick, R. F. and A. H. Levis, ''*A Mathematical Theory of Organization, Part I: Centralized Organization,*'' Polytechnic Institute of NY, Brooklyn, NY, October, 1974.

56. Duda, R. O. and P. E. Hart, *Pattern Classification and Scene Analysis*, John Wiley & Sons, 1973.

57. Feigenbaum, E. A. and J. Feldman, *Computers and Thought,* McGraw-Hill Book Company, San Francisco, CA, 1962.

58. Fikes, R. E., et. al., "Learning and Executing Generalized Robot Plans." *Artificial Intelligence,* AI-3:251–288, 1972.

59. Findeisen, Bailey, et. al., *Control and Coordination in Hierarchical Systems*, John Wiley and Sons, NY, 1980.

60. Findler, N. V., *Associative Networks. Representation and Use of Knowledge by Computers*, Academic Press, 1979.

61. Fox, M. S., "Industrial Applications of Artificial Intelligence." *Robotics,* Vol. 2: 301–311, 1986.

62. Fromm, W. and W. Eversheim, "Production Control in Highly Automated Manufacturing Systems." In *Proceedings of AUTOFACT-Europe*, Geneva, 3.1–3.13, 1982.

63. Froyd, J. and F. N. Bailey, ''Performance of Capacity Constrained Decision Makers.'' In *Proceedings of the l9th IEEE Conference on Decision and Control,* Albuquerque, NM, December 1980.

64. Fukunaga, K., *Introduction to Statistical Pattern Recognition,* Academic Press, 1972.

65. Fu K. S., ''Learning Control Systems - Review and Outlook.'' *IEEE Transactions on Automatic Control,* AC-15 (4), April 1970.

66. Fu, K. S., "Learning Control Systems and Intelligent Control Systems: An Intersection of Artificial Intelligence and Auto-

matic Control." *IEEE Transactions on Automatic Control*, AC-15 (2), Feb. 1971.

67. Fu, K. S., *Syntactic Pattern Recognition*, Academic Press, 1974.
68. Fu, K. S., *Sequential Methods in Pattern Recognition and Machine Learning*, Academic Press, 1968.
69. Fu K. S. and Booth T. L., "Grammatical Inference: Introduction and Survey - Part I." *IEEE Transactions on Systems, Man and Cybernetics*, Vol. SMC-5, No. 1, January 1975, and Part II, Vol. SMC-5, No. 4, July 1975.
70. Galbraith, J. R., *Organization Design*, Addison Wesley, Reading, MA, 1977.
71. Gallager, R. G., *Information Theory and Reliable Communication*, John Wiley and Sons, 1968.
72. Genrich, H. J., *Predicate/Transition Nets.* In Reisig W. Brauer W. and Rozenberg G., editors, *Petri Nets: Central Models and Their Properties,* Lecture Notes in Computer Science, 254, Springer Verlag, 1987.
73. Gini, M. and R. E. Smith, "Robot Tracking and Control Issues in an Intelligent Error Recovery System." In *Proceedings of the 1986 International Conference on Robotics and Automation,* Vol. 2, April 1986.
74. Goblick, T. J., "Theoretical Limitations on the Transmission of Data from Analog Sources." *IEEE Transactions on Information Theory,* 11, 558-567, 1965.
75. Goldberg, D. E., *Genetic Algorithms in Search, Optimization and Machine Learning,* Addison-Wesley, 1989.
76. Graham, J. H. and Saridis, G. N., *Linguistic Methods for Hierarchically Intelligent Control*, School of Electrical Engineering Purdue University, TR-EE 80-34, Purdue University, West Lafayette, Indiana, October 1980.
77. Graham, J. H. and Saridis, G. N., "Linguistic Decision Structures for Hierarchical Systems." *IEEE Transactions on Systems, Man and Cybernetics,* , Vol. SMC-12, No. 3, 325-333, May, 1982.

78. Green. C., "Application of Theorem Proving to Problem Solving." In *Proceedings of the 1st International Joint Conference on Artificial Intelligence,* Washington, D.C., 1969.
79. Hall, S. A., "Information Theoretic Models of Storage and Memory." *Technical Report, LIDS-TH-1232, MIT*, Cambridge, MA, August 1982.
80. Hansen, C., T. Henderson, E. Weitz and A. Mitiche, "Multisensor Knowledge Systems: Interpreting 2–D Structure." *The International Journal of Robotics Research,* Vol. 7 (No. 6), 1988.
81. Hayes-Roth, F., D. A. Waterman and D. B. Lenant, *Building Expert Systems*, Addison Wesley, 1983.
82. Hayes-Roth, F., "The Knowledge-Based Expert System: A Tutorial." *Computer Magazine*, September 1984.
83. Higashi, M. and Klir, G. J., "Measures of Uncertainty and Information Based on Possibility Distributions." *International Journal of General Systems*, 1982, Vol. 9, 43-58.
84. Hinton, G. E. and T. J. Sejnowski, "Learning and Relearning in Boltzmann Machines." In *Parallel Distributed Processing,* ed. D. E. Rumelhart and J. L. McClellan, MIT Press, 1988.
85. Horowitz, E. and S. Sahni, *Fundamentals of Data Structures*, Patomac, MD, Computer Science Press, 1976.
86. Ho, Y. C., "Team Decision Theory and Information Structures." *Proceedings of the IEEE,* Vol. 68, No. 6, June 1980.
87. Ho, Y. C. and K. C. Chu, "Team Decision Theory and Information Structures in Optimal Control Problems, Part I." *IEEE Transactions on Automatic Control,* Vol. AC-17, February 1972.
88. Ho, Y. C. and K. C. Chu, "Team Decision Theory and Information Structures in Optimal Control Problems, Part II." *IEEE Transactions on Automatic Control*, Vol. AC-17, February 1972.
89. Horaud, P., R. C. Bolles and M. J. Hannah, *Readings in Computer Vision,* Academic Press, 1989.

90. IEEE Computer Society Press, *Proceedings of the 1987 Third Conference on Artificial Intelligence Applications,* Washington, D.C., 1987.
91. "Industrial Robot Statistics." Technical Report, *International Federation of Robotics*, 1989.
92. Jain, A. K. and R. Hoffman, "Segmentation and Classification of Range Images." *IEEE Transactions on Pattern Analysis and Machine Intelligence,* PAMI-9 (5): 608–620, September 1987.
93. Jaynes, E. T., ''Information Theory and Statistical Mechanics.'' *Physical Review*, Volume 106, Number 4, March 1957.
94. Jaynes E. T., ''On the Rationale of Maximum Entropy Methods.'' *Proceedings of IEEE*, Vol. 70, No. 9, September 1982.
95. Jensen, K., "Coloured Petri Nets." In Brauer Reisig and Rozenberg (Ed.), *Petri Nets: Central Models and Their Properties:,* Lecture Notes in Computer Science, 254. Springer Verlag, NY, 1987.
96. Kalata, P. and Priemer, R., ''On Minimal Error Entropy Stochastic Approximation.'' *International Journal of Systems* SCI, 1974, Vol. 5, No. 9, 895-906
97. Kalata, P. and Priemer, R., ''When Should Smoothing Occur?.'' *Proceedings of IEEE,* September 1974.
98. Kalata, P. and Priemer, R., ''On System Identification With and Without Uncertainty.'' *Journal of Cybernetics*, 8: 31-50, 1978.
99. Kalata, P. and Priemer, R., ''Linear Prediction, Filtering and Smoothing: An Information Theoretic Approach.'' *Information Sciences*, 17, 1-14 (1979).
100. Kandel, A., *Fuzzy Mathematical Techniques with Applications,* Addison Wesley Publishing, 1986.
101. Kempf, K. G. and B. R. Fox, "Opportunistic Scheduling for Robotic Assembly." In *1985 IEEE International Conference on Robotics and Automation*, 880–889, 1985.
102. Kempf, K. G., "Manufacturing and AI." *Robotics*, Vol. 1, No. 1: 13–25, 1985.

103. Khinchin, A. I., *Mathematical Foundations of Information Theory*, Dover Publications, 1957.

104. Khinchin, A. I., *Mathematical Foundations of Statistical Mechanics*, Dover Publications, 1949.

105. Kinnucan, P., ''Computers That Think Like Experts.'' *High Technology*, January 1984.

106. Klir, G. J. and Way, E. C., ''Reconstructibility Analysis: Aims, Results, Open Problems." *Systems Research*, Vol. 2, No. 2, 141-163, 1985.

107. Kolmogorov, A. N., "On Some Asymptotic Characteristics of Completely Bounded Metric Systems," *Dokl Akad Nank, SSSR*, Vol. 108, No. 3, 385–389, 1956.

108. Koomen, C. J., ''The Entropy of Design: A Study on the Meaning of Creativity.'' *IEEE Transactions on Systems, Man, and Cybernetics*, Vol. SMC-15, No. 1, 16-30, 1985.

109. Koomen, J. A. and J. F. Allen, "Planning Using A Temporal World Model." In *Proceedings of International Joint Conference on Artificial Intelligence-8,* 1983.

110. Kusiak, A., *Intelligent Manufacturing Systems,* Prentice Hall, 1990.

111. Kusiak, A. (ed.), *Knowledge Based Systems in Manufacturing,* Taylor and Francis, 1989.

112. Lampson, B. W., M. Paul and H. Siegert, Eds. *Distributed Systems Architecture and Implementation,* Springer Verlag, New York.

113. Laurendeau, D. and D. Poussart, "3–D Model Building Using A Fast Range Finder." In *Proceedings of IEEE Computer Vision and Pattern Recognition Conference,* Miami Beach, FL, 424–427, June 1986.

114. Lautenbach, K. and H. J. Genrich, "System Modeling with High-Level Petri Nets." *Theoretical Computer Science,* Vol. 13, 1981.

115. Lawrance, J. D. and T. D. Garvey, ''Evidential Reasoning: A Developing Concept.'' *International Conference on Cybernetics and Society,* October 1982, Seattle, WA.

116. Layne, A. A., "The High Tech Needs of the Postal Services." *High Technology Magazine,* June 1989.

117. Lee, C. S. G., Gonzalez, R. C. and Fu, K. S., *Tutorial on Robotics*, IEEE Computer Society, 1986.

118. Lemmer, J. F. and Levine Kanal, *Uncertainty in Artificial Intelligence,* Elsevier Science Publishers, 1986.

119. Lennox, R. C. and D. J. Miller, "An Object-Oriented Environment for Robot System Architectures." *IEEE Control Systems Magazine,* February 1991.

120. Lifschitz, V., "On the Semantics of Strips." M. Georgeff and A. Lansky, editors, *Reasoning About Actions and Plans,* Morgan Kaufmann Publishers, 1987.

121. Ligeza, A., "Backward Versus Forward Plan Generation." In *Proceedings of the IFAC Symposium on Robot Control,* L. Basanez, G. Ferrate and G. N. Saridis, editors, 1985.

122. Lindsay, R. B. and H. Margenau, *Foundations of Physics,* Dover Publications, NY 1957.

123. Longuet-Higgins, H.C,, "A Computer Algorithm for Reconstructing a Scene from Two Projections." *Nature* (293): 133–135, 1981.

124. Macko, K., M. D. Mesarovic and Y. Takahara, *Theory of Multi-Level Systems,* Academic Press, New York, 1970.

125. March, J. G., ''Bounded Rationality, Ambiguity, and the Engineering of Choice.'' *The Bell Journal of Economics*, Autumn, 1978.

126. March, J. G. and H. A. Simons, *Organizations*, John Wiley and Sons, New York, 1958.

127. McCurthy, G. J. and K. S. Fu, "A Variable Structure Automaton Used as a Multimodal Searching Technique." *IEEE Transactions on Automatic Control,* Vol. AC-11, No. 3, July 1966.

128. McDermott, D. and E. Charniak, *Introduction to Artificial Intelligence,* Addison Wesley, 1985.

129. McDermott, D. "Reasoning About Plans." In Hobbs and Moore (ed.) *Formal Theories of the Common Sense World,* Ablex Pub., NJ, 1985.

130. McDermott, D., "A Temporal Logic for Reasoning About Processes and Plans." *Cognitive Science*, Vol. 6, 1982.

131. McInroy, J. E. and G. N. Saridis, "Reliability Based Control and Sensing Design for Intelligent Machines," in *Reliability Analysis,* ed. J. h. Graham, Elsevier, North Holland, N.Y., 1991.

132. McLaren, R. W., "A Stochastic Automaton Model for the Synthesis of Learning Systems." *IEEE Transactions on Systems Science and Cybernetics, SSC-2,* 109-114, 1966.

133. McLaren, R. W., ''Application of a Continuous-Valued Control Algorithm to the On-Line Global Optimization of Stochastic Control Systems.'' *Proceedings of the National Electronics Conference,* 25, 2-7, 1969.

134. McLaren, R. W., ''A Continuous-Valued Learning Controller for the Global Optimization of Stochastic Control Systems.'' *Seminar On Learning Processes in Control Systems,* Nagoya, Japan, 77-81, August 1970.

135. McMurtry, G. J. and Fu, K. S., ''A Variable Structure Automaton Used as a Multimodal Searching Technique.'' *IEEE Transactions on Automatic Control,* Vol AC-11, No. 3 July 1966.

136. Medioni, G. and R. Nevatia, "Segment-based Stereo Matching." *Computer Vision Graphics and Image Processing*, CVGIP-31, No. 1: 2–18, 1985.

137. Mesarovic et al., *Theory of Hierarchical, Multilevel Systems,* Mathematics in Science and Engineering Vol. 68, Academic Press, 1970."

138. Meystel, A., ''Intelligent Control of a Multiactuator System.'' *Proceedings of the 4th IFAC-IFIP Symposium on Information*

Control Problems in Manufacturing Technology, NBS, Gaithersburg, MD, Oct. 1982.

139. Meystel, A., ''Decoupling and Decentralized Control of Intelligent Multiactuator System.'' *Proceedings of the IEEE International Large Scale Systems Symposium*, October 1982.
140. Meystel, A., "Intelligent Motion Control in Authropomorphic Machines." In *Applied Artificial Intelligence*, S. Andriole, Ed. Pentrocellis Books, Princeton, NJ, 1985.
141. Meystel, A., "Intelligent Control in Robotics." *Journal of Robotic Systems*, Vol. 5, No. 4, August 1988.
142. Meystel, A. (editor), *Special Issue on Intelligent Control. Journal of Intelligent and Robotic Systems*, Vol. 2, Nos. 2–3 1989.
143. Meystel, A. M., ''Planning in the Anthropomorphical Machine Intelligence.'' *Proceedings of the International Conference on Cybernetics and Society*, October 1982.
144. Meystel, A., "Cognitive Controller for Autonomous Systems." *IEEE Workshop on Intelligent Control 1985*. 222, RPI, Troy, New York, 1985.
145. Minsky, M. L., *Artificial Intelligence*, McGraw-Hill, New York.
146. Moed, M. C., "Planning Tasks with an Emergent Connectonist/Symbolic System." *Technical Report CIRSSE-Doc. No. 46, RPI*, 1989.
147. Moed, M. C., "The Organizer: Planning Tasks with an Emergent Connectionist/Symbolic System." *Technical Report CIRSSE Document No. 42, RPI*, 1989.
148. Moed, M. C., and G. N. Saridis, "A Boltzmann Machine for the Organization of Intelligent Machines." *IEEE Transactions on Systems Man and Cybernetics*, Vol. 20, No. 5, Sept. 1990.
149. Mulgaonkar, P. G., "Multiview Image Acquisition for Postal Parcels." *Advanced Imaging*, No. 2: 44, Feb. 1991.
150. Murata, T., "Petri Nets: Properties, Analysis and Applications." In *Proceedings of the IEEE*, April 1989.

151. Narenda, K. S. and M. L. Thathachar, "Learning Automata — A Survey." *IEEE Transactions on Systems, Man and Cybernetics,* Vol. SMC-4, July 1974.

152. Narenda, K. S. and R. Viswanathan. "A Two-level System of Stochastic Automata for Periodic Random Envirnoments." *IEEE Transactions on Systems, Man and Cybernetics,* Vol. SMC-2, April 1972.

153. Nett, C.N., C.A. Jacobson, and A.T. Miller, "An integrated approach to controls and Diagnostics: The 4–parameter controller. "*Proc. 1988 American Control Conference.*

154. Nett, C. N., "Algebraic Aspects of Linear Control System Stability." *IEEE Transactions on Automatic Control,* Vol. AC-31, No. 10, October 1986.

155. Nicolic, Z. J. and K. S. Fu, "An Algorithm for Learning Without External Supervision and its Application to Learning Control Systems." *IEEE Transactions on Automatic Control,* Vol. AC-11, No. 3, July 1966.

156. Nilsson, N. J., *Learning Machines*, McGraw-Hill, 1965.

157. Nilsson, N. J. and R. E. Fikes, "Strips: A New Approach to the Application of Theorem Proving to Problem Solving." *Artificial Intelligence,* Vol. 2, 1971.

158. Nilsson, N. J., *Problem Solving Methods in Artificial Intelligence,* McGraw Hill, 1971.

159. Nilsson, N. J., "A Mobile Automation: An Application of Artificial Intelligence Techniques." *Proceedings of the International Joint Conference on AI,* Washington, D.C., 1969.

160. Nitzan, D., R. O. Duda and P. Barrett, "Use of Range and Reflectance Data to Find Planar Surfaces." *IEEE Transactions on Pattern Analysis and Machine Intelligence,* PAMI (1): 259–271, July 1979.

161. Oblow, E. M. and J. Barhen, "Mathematical Modeling and Robot Dynamics." *ORNL CESAR/TRP-03*, August 1983.

162. O'Grady, P. J., *Controlling Automated Manufacturing Systems,* Kogan Page, 1986.

163. O'Grady, P. J., et.al., "Issues in Intelligent Cell Control for FMS." *Computers in Industry,* Vol. 9:25–36, 1987.

164. Papadopoulos, N. M., ''The Gambling Decision Maker: A Model of Bounded Rationality.'' *IEEE Transactions on Automatic Control,,* Vol. AC-25, No. 3, June 1980.

165. Papoulis, A., *Probability, Random Variables, and Stochastic Processes*, McGraw Hill, 3rd Edition, 1991.

166. Parzen, E., *Stochastic Processes*, Holden-Day, 1962.

167. Passino, K. M., P. J. Antsaklis, "Fault Detection and Identification in an Intelligent Restructurable Controller." *Journal of Intelligent and Robotic Systems,* Vol. 1, 1988.

168. Pearl, J., *Heuristics,* Addison Wesley, 1984.

169. Pearl, J., *Probabilistic Reasoning in Intelligent Systems,* Morgan Kaufmann, 1988.

170. Pednault, E., *Towards A Mathematical Theory of Plan Synthesis,* Ph.D. Thesis, Dept. of EE, Stanford University, 1986.

171. Price, R., ''On Entropy Equivalence in the Time and Frequency Domains.'' *Proceedings of IRE* 43, 484-485 (1955).

172. Prigogine, I., *From Being to Becoming*, W. H. Freeman and Co., San Francisco, 1980.

173. *Proceedings of the International Workshop on Nuclear Robotics,* Tokyo, Japan, September 17-19, 1985.

174. Sacerdoti, E. D., "Planning in a Hierarchy of Abstraction Spaces." *Artificial Intelligence,* Vol. 5, No. 2, 1974.

175. Sacerdoti, E. D., *A Structure for Plans of Behavior*, Elsevier, New York, 1977.

176. Sadeh, N., K. P. Sycara, S. F. Roth and M. S. Fox, "Resource Allocation in Distributed Factory Scheduling." *IEEE Expert,* February 1991.

177. Sage, A., "Information Systems Engineering for Distributed Decision Making." *IEEE Transactions on Systems, Man and Cybernetics,* SMC-17, No. 6, 920–36, Nov./Dec. 1987.
178. Saitta, L. and A. Giordana, "Modeling Production Rules by Means of Predicate Transition Networks." *Information Sciences,* Vol. 35, 1985.
179. Sanderson, A. C., ''Parts Entropy Methods for Robotic Assembly System Design." *Proceedings of the International Conference on Robotics and Automation,* Atlanta GA, March 1984.
180. Sanderson, A. C. and H. Zhang, "Generation of Precedence Relations for Mechanical Assemblies." *Technical Report CIRSSE Document No. 34, RPI,* 1989.
181. Sanderson, A. C. and L. S. Homem de Mello, "A Correct and Complete Algorithm for the Generation of Mechanical Assembly Sequences." *Technical Report CIRSSE Document No. 33, RPI,* 1989.
182. Sanderson, A. C., and L.S. Homen de Mello, "Planning Repair Sequences Using the and/or Graph Representation of Assembly Plans." In *Proceedings of the International Conference on Robotics and Automation,* 1988.
183. Sanderson, A. C., "Applications of Neural Networks in Robotics and Automation for Manufacturing." *Technical Report CIRSSE Document No. 30, RPI,* 1988.
184. Saridis, G. N., ''Toward the Realization of Intelligent Controls.'' *Proceedings of the IEEE,* Vol. 67, 8, August 1979.
185. Saridis, G. N., *Self-Organizing Control of Stochastic Systems,* Marcel Dekker Publisher, New York, 1977.
186. Saridis, G. N., ''Control Performance as an Entropy An Integrated Theory for Intelligent Machines.'' *Report RAL19, RPI,* Troy, NY, 1983.
187. Saridis, G. N. and Stephanou, H. E., ''A Hierarchical Approach to the Control of a Prosthetic Arm.'' *IEEE Transactions on Systems, Man and Cybernetics,* Vol. SMC-7, No. 6, 407-420.

188. Saridis, G. N., ''Intelligent Robotic Control.'' *IEEE Transactions on Automatic Control*, Vol. AC–28, No. 5, May 1983.

189. Saridis, G. N. and J. Graham, ''Linguistic Decision Schemata for Intelligent Robots.'' *Preprints IFAC Congress,* Kyoto, Japan, August 1981.

190. Saridis, G. N., ''Intelligent Control-Operating Systems in Uncertain Environments.'' *Uncertainty and Control,* Ed. J. Ackermann, Springer Verlag No 70, Berlin 1985a.

191. Saridis, G. N., "Control Performance as an Entropy." *Control Theory and Advanced Technology,* 1, 2, 1985.

192. Saridis, G. N., "Foundations of Intelligent Controls." *Proceedings of IEEE Workshop on Intelligent Controls,* RPI, Troy, New York, 1985.

193. Saridis, G. N. and Valavanis, K. P., ''Information Theoretic Approach for Knowledge Engineering and Intelligent Machines.'' *Proceedings of the American Control Conference*, Boston, MA, June 1985.

194. Saridis, G. N. and H. E. Stephanou, "A Hierarchically Intelligent Control for a Bionic Arm." In *Proceedings of the Conference on Decision and Control,* Houston, TX, 1975.

195. Saridis, G. N., "Entropy Formulation for Optimal and Adaptive Control." *IEEE Transactions on Automatic Control,* Vol. 33, No. 8, 713–721, 1988.

196. Saridis, G. N., "Analytic Formulation of the IPDI for Intelligent Machines." *Automatica, the IFAC Journal*, 25, No. 3, 461–467, 1989.

197. Saridis, G. N. and F. Wang, "Structural Formulation of Plan Generation for Intelligent Machines." *Technical Report CIRSSE Document No. 40, RPI,* 1989.

198. Saridis, G. N. and J. E. McInroy, "Reliability Analysis in Intelligent Machines." *Technical Report CIRSSE Document No. 39, RPI*, 1989.

199. Saridis, G. N. and M. C. Moed, "Analytic Formulation of Intelligent Machines as Neural Nets." *IEEE International Symposium on Intelligent Control*, Washington, D.C., August 1988.

200. Saridis, G. N. and H. E. Stephanou, "A Hierarchical Approahc to the Control of a Prosthetic Arm." *IEEE Transactions on Systems, Man and Cybernetics*, Vol. SMC-7, No. 6, 407–420, 1977.

201. Saridis, G. N. and K. P. Valavanis, "Analytical Design of Intelligent Machines." *Automatica the IFAC Journal*, Vol. 24, No. 2 123–133, 1988.

202. Saridis, G. N., "Architectures for Intelligent Machines." *CIRSSE Technical Report No. 96*, RPI, August 1991.

203. Seetharaman, G. S. and T. J. Young, *Advances in Computer Vision,* Chapter A, Region Based Approach to 3–D Motion Analysis. (Ed.), Thomas Huang, JAI Press, 1989.

204. Seetharaman, G. S. and T. J. Young, *Image Understanding in Unstructured Environment,* Volume 2 of *Series in Automation,* Chapter A Region Correspondence Approach to the Recovery of 3–D Motion and Structure in Dynamic Scenes, pp. 75–124. World Scientific, New Jersey, 1989.

205. Sen, P. and R. F. Drenick, "Information Processing in Man-Machine Systems." *Proceedings of the Fourth MIT-ONR Workshop on C3 Systems, LIDS-R-1159*, MIT, Cambridge, MA, October 1981.

206. Shafer, G., *A Mathematical Theory of Evidence*, Princeton University Press, Princeton, NJ, 1976.

207. Shannon, C. and Weaver, W., *The Mathematical Theory of Communication*, Illinois Books, 1963.

208. Shannon, C. E., "Prediction and Entropy of Printed English." *Bell Systems Technical Journal 30*, 1951: 50- 64.

209. Shanthikumar, J. G. and J. A. Buzacott, "Models for Understanding FMS." *American Institute Industrial Engineers Transactions,* No. 12: 339–349, 1980.

210. Sheridan, T. B. and W. R. Ferrell, *Man-Machine Systems*, MIT Press, Cambridge, MA, 1974.

211. Shirai, Yoshiaki, *Three Dimensional Computer Vision*, Springer-Verlag, New York, 1987.

212. Shrikhande, N. and G. Stockman, "Surface Orientation from a Projection Grid." *IEEE Transactions on Pattern Analysis and Machine Intelligence*, PAMI 11: 650–655, June 1989.

213. Simon, H. A., "A Behavioral Model of Rational Choice." *Quarterly Journal of Economics*, Vol. 69, 1956.

214. Simon, H. A., *Administrative Behavior*, MacMillan Co., NY, 1961.

215. Singh, M. G. and K. Hindi, "A Multilevel Multilayer Framework for Manufacturing Control." *Journal of Intelligent Robotic System*, Vol. 4, No. 1, 1991.

216. Solomon A. D., "World Model." *ORNL CESAR/TRP-02*, August 1983.

217. Spong, M.W. and M. Vidyasagar, "Robust Linear Compensator Design for Nonlinear Robot Control." *IEEE Journal Robotics Automat*, RA-3(4) (1987).

218. Stabile, D. A., "*The Design of Information Structures: Basic Allocation Strategies for Organization.*" M.S. Thesis, LIDS-TH-1098, MIT, Cambridge, MA, June 1981.

219. Stabile, D. A., A. H. Levis and A. Hall, "Information Structures for Single Echelon Organizations." *LIDS-P 1180, MIT,* Cambridge, MA, January 1982.

220. Stark, H. and Woods, J., *Probability, Random Processes and Estimation Theory for Engineers*, Prentice Hall, 1986.

221. Stefik, M., "Planning with Constraints" (Molgen: Part 1). *Artificial Intelligence*, Vol. 15, No. 2, 1981.

222. Stefik, M., "Planning and Meta Planning" (Molgen: Part 2). *Artificial Intelligence*, Vol. 15, No. 2, 1981.

223. Stellakis, H. M., *Design of A Fuzzy Logic Based Organizer for an Intelligent Robotic System.* M.Sc. Thesis, Northeastern University, Boston, MA 1989.

224. Stephanou, H. E. and Saridis, G. N., *Hierarchical Intelligent Control of a Prosthetic Arm*, School of Electrical Engineering Purdue University, TR-EE 76-8, July 1976.

225. Stephanou, H. E. and Lu, S. Y., "Measuring Concensus Effectiveness by a Generalized Entropy Criterion." *Proceedings of the Conference on Artificial Intelligence*, Denver, CO, 1984.

226. Stephanou, H. E. and A. M. Erkmen, "Model-Based Sensory Pattern Fusion." *Journal of Robotic Systems*, Vol. 7, No. 3, June, 1990.

227. Triendl, E. E., T. C. Henderson and R. Winter, "Edge and Shape-based Geometric Registration." *IEEE Transactions On Geoscience and Remote Sensing*, GE-23, No. 3: 334–342, May 1985.

228. Tsai, R. Y. and T. S. Huang, "Uniqueness and Estimation of Three Dimensional Motion Parameters of Rigid Objects with Curved Surfaces." *IEEE Transactions on Pattern Analysis and Machine Intelligence,* PAMI 6, 545–554, 1985.

229. Tsetlin, M. L., "On the Behavior of Finite Automata in a Random Media." *Automation and Remote Control,* Vol. 22, No. 10, October 1961.

230. Tzafestas, S. G., (editor), *Intelligent Robotic Systems,* Marcel Dekker, 1991.

231. Valavanis, K. P. and G. N. Saridis, "Mathematical Formulation for the Organization Level of Intelligent Machines." *Proceedings of the IEEE International Conference on Robotics and Automation*, San Francisco CA, April 1986.

232. Valavanis, K. P. and G. N. Saridis, "Information Theoretic Modeling of Intelligent Robotic Systems." *IEEE Transactions on Symstems, Man and Cybernetics,* Vol. 18, No. 6, 1988.

233. Valavanis, K. P. and Saridis G. N., "Analytical Design of Intelligent Machines." *Proceedings of the SYROCO Conference*, Spain, November 6-8, 1985.

234. Valavanis, K. P., "*On the Real-Time Control of a VAX-11/750 Computer-Controlled PUMA-600 Robot ARM.*" M. Sc. Thesis, RPI, Troy NY, June 1984.

235. Valavanis, K. P. and P. H. Yuan, "Hardware and Software for Intelligent Robotic Systems." *Journal of Intelligent and Robotic Systems,* Vol. 1, No. 4, 1989.

236. Valavanis, K. P. and S. J. Carelo, "An Efficient Planning Technique for Robotic Assemblies and Intelligent Robotic Systems." *Journal of Intelligent and Robotic Systems,* Vol. 3, No. 4, 1990.

237. Valavanis, K. P., *A Mathematical Formulation for the Analytical Design of Intelligent Machines.* Ph.D Thesis Dissertation, Rensselaer Polytechnic Institute, Troy, N.Y., 1986.

238. Valavanis, K. P., Gauch, G., and J. Zheng, "On Impulse Noise Removal in Color Images." In *Proceedings of the 1991 IEEE International Conference on Robotics and Automation,* Sacramento, CA, April, 1991.

239. Valavanis, K. P., C. A. Jacobson, and B. Gold, "Integrated Control and Failure Detection with Application to the Robot Payload Variation Problem." *Journal of Intelligent and Robotic Systems,* Vol. 4, No. 2, 1991.

240. Valavanis, K. P., "On the Hierarchical Modeling Analysis and Simulation of FMS with Extended Petri Nets." *IEEE Transactions on Systems, Man and Cybernetics,* Vol. 20, No. 1 , 1990.

241. Valavanis, K. P. and G. N. Saridis, "A Review of Intelligent Control-based Methodologies for Modeling and Analysis of Hierarchically Intelligent Systems." *Proceedings of the IEEE International Symposium on Intelligent Controls,* Philadelphia, PA, Sept. 1990.

242. Valdes-Perez, R. E. and Conant, R. C., "Information Loss Due to Data Quantization in Reconstructibility Analysis." *International*

Journal and General Systems, 9 (1983), 235-247.

243. Vamos, T., "Metalanguages — Conceptual Models, Bridge Between Machine and Human Intelligence." *Working Paper E/37, Hungarian Academy of Science*, 1986.
244. Van Campenhout, J. M. and Cover, T. M., "Maximum Entropy and Conditional Probability." *IEEE Transactions on Information Theory,* Vol. IT-27, No. 4, July 1981.
245. Van Trees, H. L., *Detection, Estimation and Modulation Theory,* John Wiley, 1968.
246. Vidyasagar, M., *Control System Synthesis.* MIT Press, Cambridge, Mass. (1985).
247. Viterbi, A. J., "On the Minimum Mean Square Error Resulting From Linear Filtering of Stationary Signals in White Noise." *IEEE Transactions on Information Theory,* 594-595, 1965.
248. Waltz, M. D. and K. S. Fu, "A Heuristic Approach to Reinforcement Learning Control Systems." *IEEE Transactions on Automatic Control, AC-10,* Vol. 10, No. 4, October 1965.
249. Wang, F. Y., "*A Coordination Theory for Intelligent Machines.*" Ph.D. Thesis, *Technical Report CIRSSE No. 59, RPI,* 1990.
250. Wang, F. Y. and G. N. Saridis, "Task Plan Generation for Intelligent Machines." *Proceedings of the IEEE International Symposium on Intelligent Controls*, Philadelphia, PA, Sept. 1990.
251. Wang, F. and G. N. Saridis, "A Coordination Theory for Intelligent Machines." *Automatica, The IFAC Journal*, Vol. 35, No. 5, Sept. 1990, 833–844.
252. Wang, F. Y., K. J. Kyriakopoulos, A. Tsolkas, and G. N. Saridis, "A Petri Net Coordination Model of Intelligent Mobile Robots." *IEEE Transactions on Systems, Man, and Cybernetics,* Vol. SMC–21, No. 4, July 1991.
253. Wang, F. Y. and G. N. Saridis, "Coordination Structures for Specification of Integration in Intelligent Machines." *Proceedings of the 1991 Conference on Robotics and Automation*, Sacramento, CA, April 1991.

254. Wang, F. and G. N. Saridis, "A Model for Coordination of Intelligent Machines Using Petri Nets." *IEEE International Symposium on Intelligent Control*, Washington, D.C., August 1988.

255. Webster's New Collegiate Dictionary, Merriam.

256. Weidemann, H. L., "Entropy Analysis of Feedback Control Systems." In *Advances in Control Systems*, Vol. 7, 1969, C. Leondes edt. Academic Press.

257. Wiener, N., *Cybernetics*, MIT Press, MA, 1962.

258. Wilkins, D. E., "Domain-independent Planning: Representation and Plan Generation." *Artificial Intelligence,* Vol. 22, No. 3, 1984.

259. Wilkins, D. E., "Recovering from Execution Errors in Sipe." *Computational Intelligence,* Vol. 1, No. 1, 1985.

260. Winston, P., *Artificial Intelligence*, Addison-Wesley, New York.

261. Wohl, J. G., "Battle Management Decisions in Air Force Tactical Command and Control." *Report M79-233, MITRE,* Bedford, MA, December 1979.

262. Wong A. K. C. and You, M., "Entropy and Distance of Random Graphs with Application to Structural Pattern Recognition." *IEEE Transactions of Pattern Analysis and Machine Intelligence*, Vol PAMI-7 No. 5, September 1985.

263. Yager, T., Ovchinnicov and Nguyen, *Fuzzy Sets and Applications,* Selected papers by L. A. Zadeh, Willey-Interscience Publishing, 1987.

264. Yuan, P.H., "*Design of an Intelligent Robotic System Organizer Via Expert Systems Techniques.*" M.S. Thesis, Northeastern University, Boston, MA, 1988.

265. Zadeh, L. A., "Fuzzy Sets as a Basis for a Theory of Possibility." *Fuzzy Sets and Systems* Vol. 1, 3-28, (1978).

266. Zadeh, L. A., "Fuzzy Sets." *Information and Control*, Vol. 8, 338-353, 1965.

267. Zames, G., "On the Metric Complexity of Casual Linear Systems, ε—entropy and ε—dimension for Continuous Time." *IEEE*

Transactions on Automatic Control, Vol. AC-124, 220–230, April, 1979.

268. Zheng, Y. F., "Integration of Multiple Sensors into a Robotic System and its Performance Evaluation." *IEEE Transactions on Robotics and Automation*, Vol. 5, No. 5, October 1989.
269. Zheng, J., Gauch, G., and K. P. Valavanis, "Object Extraction from Color Robotic Vision Systems." In *Proceedings of the 6th IEEE International Symposium on Intelligent Controls,* Arlington, Va, August, 1991.
270. Zimmermann, H. J., *Fuzzy Set Theory and Its Applications,* Kluwer-Nijhoff Publishing, 1985.
271. Zimmermann, H. J., *Fuzzy Sets, Decision Making, and Expert Systems,* Addison-Wesley Publishing, 1985.

Appendix B GLOSSARY

$p(.)$:	Probability Density Function (pdf)
$p(.,.)$:	Joint pdf
$p(./.)$:	Conditional pdf
$H(.)$:	Entropy Function
$H(.,.)$:	Joint Entropy
$H(./.)$:	Conditional Entropy
$\overline{H}(.)$:	Entropy Rate
$T(. : .)$:	Transmission of Information
$\overline{T}(. : .)$:	Transmission Rate
τ :	Interarrival time
G :	System
G_i:	Subsystem i
G_{int}:	Internal variables of G
G_{out}:	External variables of G
F :	Total Acitivity Rate of G
F_t:	Throughput Rate of G
F_b:	Blockage Rate of G
F_c:	Coordination Rate of G
F_n:	Noise Rate of G
Ω_s:	Space of Information Signals/Knowledge
Ω_x:	State Space
Ω_u:	Space of Admissible Controls

T :	Temperature
S :	Boltzmann's Entropy
y :	Gibbs Energy
k :	Boltzmann's Constant
E	Total Energy
L :	Lagrangian
$\overline{L}$:	Average Lagrangian
$x(t)$:	System States
$u^*(x,t)$:	Optimal Control Input
$V(x_o,u,t)$:	Generalized Energy Function
$E\{\ \}$:	Expected Value
s_i:	State of System
K_{ij}:	Knowledge Between two states s_i, s_j
w_{ij}	State of Transition Coefficients
K_i:	Total Knowledge of State i
R_{ij}:	Flow of Knowledge Between States i, j
R_i:	Total Flow of Knowledge from state i
$C_1 = \{c_1, ..., c_M\}$:	Set of Input Commands
$U = \{u_1, ..., u_M\}$:	Set of Compiled Commands
$E_t = \{e_1, ..., e_N\}$:	Set of Primitive Events
x_i:	Binary Random Variable Associated with e_i
A_{jm}:	mth Activity Related to jth Compiled Command
X_{jm}:	Binary Vector Corresponding to A_{jm}:
B_{jmr}:	Augmented Activity

Y_{jmr}:	Augmented Binary Vector Corresponding to B_{jmr}:
M_{jmr}:	Permutation Mask
Z_{jmr}:	Set of Complete Plans
$\lambda(.)$:	Symmetrical Loss Function
α, β, γ :	Euler Angles
q_1, q_2, $q_3 \ldots$:	Generalized Coordinates
r_x, r_y, r_z :	Cartesian Vector Components
$g(.)$:	Discriminant Function
σ:	Variance
v_n:	Signal of Constant Value
$R(./.)$:	Conditional Risk Function
H_{QP}:	Object Positional Entropy
H_{QO}:	Object Orientational Entropy
$J_1, J_2, \ldots$:	Performance Indices, Cost Functions

Appendix C ABBREVIATIONS

IM:	Intelligent Machine
IRS:	Intelligent Robotic System
RA:	Robotic Assembly
RSO:	Robotic System Organizer
IROS:	Intelligent Robot Operating System
FMS:	Flexible Manufacturing System
AI:	Artificial Intelligence
DAI:	Distributed Artificial Intelligence
MI:	Machine Intelligence
DMI:	Distributed Machine Intelligence
PN:	Petri Net
PNT:	Petri Net Transducer
IPDI:	Increasing Precision with Decreasing Intelligence
PLIR:	Partition Law of Information Rates
GPLIR:	Generalized Partition Law of Information Rates
DM:	Decision Maker
SA:	Situation Assessment
RS:	Response Selection
IF:	Information Fusion
CI:	Command Interpretation
K:	Total Machine Knowledge
R:	Rate of Machine Knowledge
DB:	Data Base

MR:	Machine Reasoning
MP:	Machine Planning
MDM:	Machine Decision Making
MFB:	Machine Feedback
MME:	Machine Memory Exchange
KB:	Knowledge Base
IE	Inference Engine
VSC:	Vision System Coordinator
SSC:	Sensor System Coordinator
MSC:	Motion System Coordinator
GSC:	Gripper System Coordinator
PWR:	Pressurized Water Reactor
BWR:	Boiled Water Reactor
SVi:	Safety Valve i
RVi:	Relief Valve i
IVi:	Isolation Valve i
FLi:	Flange i
Pi:	Pipe i
CTi:	ith Command Type Table
ETi:	ith Instruction Type Table
IST:	Instruction Set Table

INDEX